In Our Blood

In Our Blood
Four Coal Mining Families

By Matt Witt
Photographs by Earl Dotter

Highlander Research and Education Center

Printed in the U.S.A. First edition

Library of Congress Catalog Card Number: 78-71518
ISBN 0-9602226-1-8

All proceeds from the sale of this book will go to a special fund to support non-profit education projects to benefit working people in the American coalfields. The fund will be administered by the non-profit Highlander Research and Education Center in New Market, Tennessee. Highlander has worked closely with labor unions, civil rights groups, and other workers' organizations since the 1930s.

To order copies of this book, send a check or money order for $6.95 each to Highlander Center, P.O. Box 32313, Washington, D.C. 20007. Please allow four to six weeks for delivery.

Introduction

Without coal miners, America would grind to a halt. It would no longer be possible to produce the steel our economy needs to function. In many parts of the country, the lights would go out as electric utilities shut down. Homes would be without heat; factories would be without power.

Americans depend on coal miners, and will depend on them more as other sources of energy become too scarce or unreliable. Yet most people know little about life in the coalfields. News from coal communities seeps out only when an underground explosion takes a large number of miners' lives, or when a strike threatens to cut off the nation's coal supply. Once the "crisis" passes, business as usual resumes, and mining families are again forgotten.

Although most people would not work in a coal mine for any amount of money, two hundred thousand American men and women risk their lives each day to bring coal out of the ground.

They live in the rolling hills of Pennsylvania, the isolated hollows of southern Appalachia, the flat farm country of the Midwest, and the mountains and deserts of the West. They are the sons and daughters of European immigrants, of black field hands from the South, of Navajo Indians.

Despite their regional and cultural differences, coal miners everywhere have much in common. They face constant danger on the job. Their schools, housing, and health care facilities are generally among the worst in the nation. To improve conditions in their communities, they have had to struggle together for generations against the coal companies, and at times against weak or corrupt union leaders.

The four mining families portrayed in this book were chosen because they came from a variety of backgrounds, and because they and their neighbors have been involved in organized efforts to bring about changes in coalfield life. While some mining regions and ethnic groups are not represented, these profiles help tell the story of all coal miners. It is a story that could not have been told without the generous cooperation of the families, who opened their lives to us even though they were puzzled by our interest.

"Why would anybody want to write a book about me and my family?" asked one of the miners, John Socoski. "Books aren't supposed to be about people like us. We're not that important."

We think they are.

M.W.
E.D.

The Promised Land

A narrow shaft of light breaks the tunnel's darkness, reflecting off the dust that swirls in the cold mine air. Aiming the light clipped to his hard hat so it will shine into his tool box, John Socoski is able to see the wrench he needs. He picks it up, shifts his light onto the huge mining machine next to him, and reaches across to tighten the bolts on the machine's control box cover.

''The company runs this equipment into the ground and then they can't figure out why it breaks down,'' he says. ''They try to get another day or two out of it, and let the mechanics take care of it on Saturday, but it doesn't always make it that long.''

The cover fastened, he slides off the machine to take a break. Standing with his back and neck bent because the underground tunnel is only forty-five inches tall, he wipes his forehead with the sleeve of his overalls, adding to the coal and grease which covers his face.

''Know what I'm going to do some day to get away from all this?'' he says. ''I've been saving up to take Shirley to Hawaii for our twenty-fifth anniversary. Hawaii, can you beat that? We'll take one of these tours where it's all arranged for you. I can't see hitchhiking through the back alleys, can you?''

The machine operator looks over at John and asks through a wad of

JOHN SOCOSKI WORKS ON A MINE VEHICLE

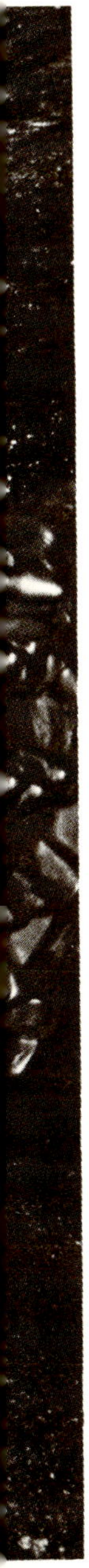

chewing tobacco, "Hey, did you find the problem yet?"

"Yeah," John answers. "Let's start it up again, and see what happens."

The operator pulls a lever, and more than four hundred volts of power surge into the "continuous miner." The thirty-foot-long machine begins to roar, then suddenly lurches forward toward the uncut coal. Thick rotating bits tear into the shiny black seam, making the air heavy with dust. Powerful enough to cut and load tons of coal in minutes, the machine will reach out for fingers, arms, and legs as well, camouflaging its attacks with deafening noise. As it cuts the coal, pockets of methane gas are released into the mine atmosphere. If gas or coal dust are allowed to build up, all it will take is one small spark to ignite them. Like other miners, John carries a "self-rescuer" cannister on his belt which will help him breathe for the first hour after an explosion. But if disaster strikes, he will probably never get a chance to use it. Explosions and fire rocket instantly through a coal mine with the force of a bomb blast. There is no time to run and nowhere to hide.

As John watches the machine load coal onto waiting shuttle cars, layers of coal peel off the walls and fall into an ankle-deep stream created on the mine floor by underground springs. The roof groans overhead as the weight of the earth shifts where coal has been removed. Steel and wooden supports are supposed to hold the roof up, but every year thousands of miners are injured and dozens are killed in rock falls. About four years ago, John helped an experienced miner named Harry Koptchak get a job here at the Rushton mine. Several months later, Koptchak was crushed without warning by a falling piece of slate.

At age forty-one, John has already lost the use of his right eye in a mine accident. If he works another twenty-five years, he can expect to suffer several other serious injuries. On the average, one American miner is killed every other day, and one is injured every ten minutes. Anyone who beats those odds has a three-out-of-four chance of spending retirement suffering from black lung disease. Coal dust clogs the miners' lungs, slowly taking away their breath. Once black lung takes hold, there is no cure.

Satisfied that the machine is fixed, John turns to pick up his tools. "Yeah, I'm looking forward to taking my wife to Hawaii for our

twenty-fifth,'' he sighs. ''You never know if you'll be around for the twenty-sixth.''

In the evening after work, John heads to the Slovak Club for a few drinks. At one end of the bar, two middle-aged coal miners' wives are singing along with the country tunes from the jukebox. At the other end, Harry Koptchak's son, Buck, is playing darts. He started working at the Rushton mine a few months after his father was killed, because his family needed the income and he didn't know of any other well-paying jobs in the area.

''Buck,'' John says, ''you better be at our next local union neeting. We're going to have to decide what to do.''

Buck finishes a turn with the darts and hands them to John. As John shoots, the two men argue about the experiment that has been going on at the Rushton mine for several years. Under an agreement between the company and the local union of the United Mine Workers of America, some of the miners are allowed to direct their own work. The miners hoped the arrangement would give them the power to protect their health and safety on the job, while the company hoped it would increase productivity. The only such experiment ever attempted in an American coal mine, it has been plagued by a series of disagreements between the miners and management. John helped set up the program, but now thinks the UMWA should withdraw its participation. Buck isn't so sure.

''Your turn,'' John says, as he sits back down and asks Ruthie, the bartender, for another can of tomato juice. He pulls off the tab and pours the contents into his beer, making the pale pink mixture which is his favorite drink.

''I'm telling you, Buck,'' he says, ''that program might have worked before, but it's not working now. The company's just not keeping its part of the bargain.''

Another round of darts, more beers, and soon the discussion between John and Buck is reduced to good-natured repetition. Ruthie signals that it is closing time; the debate will have to be finished at the next union meeting.

Driving home, John points to a well-kept, white, two-story frame house.

''That's my homestead,'' he says. ''That's where I was raised. Born and raised right here in Osceola Mills. They used to say, 'Osceola Mills: the town with fourteen bars and fourteen churches.'

''If my dad was still alive, you could ask him about the old days in this town. He put in more than forty years in the mines right around here. In fact, he lost his right eye like I did, in an accident.

''My mom still lives in that house,'' he says. ''You talk to her. She can tell you what things used to be like in this part of Pennsylvania, before we had the union.''

''I guess my family and my husband's family thought it would be milk and honey when they came to America,'' says John's sixty-nine-year-old mother, Anna Socoski. ''They had hard lives in Czechoslovakia, and people told them stories about how it would be the promised land over here. They weren't prepared for what they found.''

Anna's grandfather, Michael Petrik, Sr., came to Pennsylvania in the 1890s, leaving his family in the old country until he could make enough money to send for them. Landing in New York City, he was

PHOTOS BY LEWIS HINE

directed straight to the coalfields of the Moshannon Valley, where immigrant labor was welcome.

Coal operators there were locked in a running battle with the UMWA, which had been organizing in Pennsylvania since the union's founding convention in 1890. The companies discovered that it was easier to keep the union out with a work force of immigrants who spoke many different languages and couldn't afford to lose their jobs. Agents were sent to the docks of New York City to lure newcomers with promises of big money and easy living, and soon the Pennsylvania countryside was filled with villages of Slovaks, Hungarians, Italians, Poles, Lithuanians, and Russians.

Michael Petrik, Jr., was lucky that his father was able to send for him. Many of the men who came to America never saved enough money to pay their families' passage; years would go by, the women who stayed behind no longer wives but not really widows, the men assuring themselves until their last days that next month or next year they would arrange a reunion.

"My father told me they were treated like animals on the boat he came on," Anna recalls. "I guess in a lot of ways it was like the ships that brought slaves from Africa. They kept all the people in one room below deck, and all they had to eat was maybe a little bread or a common pot of soup. They had to eat fast because there were so many rats swarming around, trying to get the food. When he got to New York, they kept him locked up for a week or two so they could take the lice off the people. Some of them were sent right back home because they had gotten diseases during the trip."

Immigration officials improvised as they translated the Slovak names, changing "Petrik" to the more familiar "Patrick." A few years later, when John's father and uncle came over on separate ships after their parents both died young, one was recorded as Stephen Socoski, the other as Joseph Sakosky.

Anna's father had to pay a coal company five dollars to give him a job in the mines. The fee bought him the right to work a "place," or room underground. First, he would use dynamite to loosen a block of coal. Then, on his knees, he would shovel it into cars. He was paid by the ton, with the company controlling the scales, so that what looked to him like two tons often was recorded as only one or one-and-a-half.

On their own time, the Patricks and Socoskis had to load out tons of worthless rock that came loose with the coal, and set timbers to support the roof. They got up at three a.m. to walk to the mine entrance and then several miles inside, and returned home at four or five o'clock in the afternoon. There were days they didn't earn a dime because so much clean-up work had to be done. When some of their fellow workers complained about the "clean-up system," they were given the places with the most rock and thinnest coal seams.

Without a union, the miners had no way to force the companies to spend money on health and safety precautions. Open flames were used for light, and fires and explosions were frequent; Anna's grandfather was killed when flames ignited a supply of carbide used to fuel the miners' lights. Because the companies refused to pump out the underground rooms, miners had to kneel in water for ten or twelve hours a day. In winter, they walked home with ice clinging to their legs. Stephen Socoski later suffered from rheumatism as a result. The tracks on which coal cars traveled in and out of the mine were not maintained; once, when Anna's father was driving a load of coal to the outside, the cars hit a rough spot and his brakeman was crushed against the roof. Young boys who started working around the mines at age ten or eleven often lost fingers or limbs by the time they were teenagers.

"As long as there were plenty of people who needed jobs, the company didn't worry about it if you got hurt," Anna says. "They just got rid of you and found somebody else to put in your place."

Anna was the oldest of twelve children. She remembers getting up at six a.m. to carry the family's water from a nearby stream and to feed and milk the cows. There was no school in the coal camp where the Patricks lived; on days when there wasn't too much rain or snow, she would walk to the school in Osceola Mills four miles away. It would usually be dark by the time she came home. By the light of a kerosene lantern, she would help her mother make dinner and then heat water and

wash dishes. These were formidable chores since, in addition to the large family, her parents kept as boarders a group of Italian miners who liked to eat large quantities of cooked-to-order spaghetti.

"You would have thought you were in a little town in Europe," she recalls. "We were too poor to have a church building, but people still had their special events, their christenings and weddings and holidays. I can remember my Uncle John made a fiddle out of a cigar box, and he used to play while the others would dance.

"The toys we had, people would make them," she says. "Like I remember when Dad made my little brother a play sled out of wood. Well, then my brother came down with whooping cough, and my mother had to walk miles and miles, carrying her three-year-old boy, to find a doctor. The doctor just gave her some black medicine that I guess they gave you at that time no matter what was wrong with you.

"Well, he moaned in his crib for about a week, and then he died. And my dad, he grabbed that little wooden sled he had made for him, and he stuffed it in the cookstove to burn. He didn't want to be reminded."

One of Anna's early memories is the hunger in her stomach as the smell of bacon wafted over the camp from the quarters of the Coal and Iron Police, professional strikebreakers paid by the state and the company. By the end of World War I, many of the Moshannon Valley mines had been organized by the UMWA. But when the demand for coal dropped after the war and fewer miners were needed, many companies gave their men a choice of working without a union for lower wages or not working at all. Thousands of miners went on strike, but the operators had no trouble finding replacements.

'The men couldn't stop the scabs from taking their jobs, because the company had the law on their side and the guns," Anna recalls. "Each day the men would try to block the way to the mine but the police would escort the scabs through. So finally the women got into it, picketing with brooms, beating on washtubs, anything they could do to raise a fuss. But it didn't do any good. The union was out again, and the company could do as they pleased."

In 1934, a few years after Anna and Stephen Socoski were married, they moved from a camp called Low Coal to the house in Osceola Mills where John was born. Living in town, they could have running water and electricity, easier access to stores and schools, and a chance to give their five sons and two daughters a proper upbringing in the Russian Orthodox Church.

"My mom used religion to teach us to do the right thing," John recalls. "Me and my brother, Al, we always had to be scrounging for a dollar because my dad was sick, and the way the bottom dropped out for coal mining around here, you couldn't find a job. We sold rugs that my mom made on a handloom, and we used to go down to the dump and find junk, like iron and stuff, and drag it home to sell.

"So this one time we were down at the dump, we saw these envelopes burning, and Al sees that one of them has a five-dollar bill sticking out of it. We start looking around, and these are church envelopes, from the Roman Catholic church, and a lot of them still have money in them.

"We went through and we found about sixty bucks. We took it home to give to my mom, and we were prouder than peacocks—sixty bucks! And we said, 'We're going to keep it. They don't want the money. They're burning it.'

"Well, we both got a whipping for wanting to keep that money, and I had to take it to the priest. My mom said I was gonna go to Hell, see, for keeping that money.

"She was right, that money didn't belong to us," he says. "But we were sure disappointed that day."

It takes two cars to transport John and Shirley and their seven kids to St. Mary's Russian Orthodox Church on Sunday. In the past ten years, the priest has stopped conducting the service in Russian, and the men and women no longer sit separately, but otherwise the rituals of past centuries are preserved.

Wearing ornate, formal robes, the priest repeats traditional chants

MINE STRIKE IN OSCEOLA MILLS, EARLY 1920s

while the voices of the choir reverberate in the tiny building. Several times during the service, an older man pulls on a rope to ring the church bell. Nick, the Socoskis' second oldest, helps the priest give communion, which is taken by some of the older folks and most of the toddlers, including the Socoskis' three-year-old, Nate.

Every Sunday the experience is the same: the near-magic mood created by the singing, the reassuring repetition of Jesus' teachings on how to live, the feeling of solidarity with the neighbors. While the Socoskis' two daughters, ages six and eleven, enjoy going to church, the four older boys seem to attend mainly to please their parents. Either way, the ritual helps keep family and community together.

After church, the family sits down to the Sunday dinner Shirley has prepared: "holupka" (cabbage stuffed with ground beef cooked in tomato sauce), cauliflower, home-grown beets, mashed potatoes and gravy, and white bread. Except for an occasional appointment in the beauty shop she runs in their home, Shirley's days are a continuous cycle of washing dishes, doing laundry, putting away food, shopping, cleaning, and taking care of Nate. Keeping up with the housework is like trying to empty the ocean with a bucket, but she says she chose her job, and seems to enjoy it. Before she married John, she was enrolled at Penn State with hopes of becoming a doctor, but she gave up her education to raise her family.

"I always agreed with what John's father used to say: 'No matter how many kids you've got, if you're lucky enough to have another one, there's always a place for them,'" she says. "I think the first one and the last one are always the best, though. The first one because they're so special to you, and the last one because you finally have enough help around that you can relax a little."

The dinner conversation turns to the boys' rock-and-roll band, "The Last Resort." Their next engagement will be at the YMCA. Three of them are in the band, and a fourth handles the light show. When they play for the teenagers at the Y, they turn up their amplifiers as loud as they go and scream into the microphones. For the older folks at the American Legion hall, they shift to polkas. Playing both types of music well, they have made enough money to buy top-quality equipment so they can compete with the bands started by the bosses' kids—"the richies," as they call them.

HOHNER

"My kids don't give me no trouble," John says. "I don't have no cops knocking on my door. They know they have to help around the house, like with bringing in the hay for the cows we keep. But we give them their recreation too. We don't like to keep them cooped up.

"I give them respect, and they do the same," he says. "When they're eighteen, I give 'em a party, let 'em bring their friends over, have all the beer they want. After that they don't drink much; they don't care about it."

The three oldest boys have all been starters on the varsity football team at Philipsburg-Osceola High.

"I wanted my boys to play football," says John, who once had to turn down a two-year football scholarship to the University of Delaware so he could work to support his parents. "I learned about unions from my mom and dad and in the first jobs I had, but believe it or not, the way I really learned about teamwork was from football. That's how I learned that how strong you are depends on whether you stick together. I wanted them to learn that."

John got his first job in a coal mine in 1964, when the Rushton mine opened. Warren Hinks, the company president, made it clear to the men he hired that Rushton would be a non-union operation, and John and others from UMWA families played along until they were on the payroll. They they started holding secret meetings with union organizers, and before long, a government-supervised election was arranged to determine if a majority of the workers wanted UMWA representation.

Hinks spent long hours underground, day and night, warning against a pro-union vote. Rumors were circulated that wages would be lower under a UMWA contract, that union supporters would be laid off, that the mine might not be able to operate at all with a union.

The union narrowly won the election, giving it the right to bargain for a contract covering the mine, but Hinks had no intention of signing up. During a month-long strike, he personally led strikebreakers across the picket line. On payday, those who were working would wave their checks at the pickets to discourage them.

Then Joseph (Jock) Yablonski, a UMWA executive board member, began visiting John and other Rushton miners at their homes to discuss ways to put more pressure on the company. Within a few days, other mines in central Pennsylvania which supplied coal to the same electric company as Rushton were shut down by pickets. The strike became too great an inconvenience for the power company, Hinks' only customer, and soon he was forced to accept the union.

A contract was signed, but there was still no peace at Rushton. The miners' main concern was the company's heavy-handed emphasis on productivity.

"Everything my dad had told be about the coal mines was happening to me too," John recalls. "This company was just like the ones he worked for. Profits, that's the only thing that meant anything to them."

At best, a coal mine is a hostile environment, and it can be made even more dangerous if miners are forced to take shortcuts. They may be required to run their machines so fast that they cannot see other workers in their path. Necessary maintenance may be put off, forcing miners to run equipment with bad brakes, for example, or to work with worn-out electric cables that should be replaced instead of patched. To save time, miners installing roof supports may be told to use whatever materials are

on hand, even if they are not the right size, or to by-pass the step of putting up temporary props for protection while they work. Crews may be asked to work short-handed, or with unqualified personnel in key jobs.

The Rushton miners knew that if they protested unsafe practices, the company could find excuses to fire them, deny them an equal chance at high-paying overtime work, or assign them unpleasant jobs. Many workers at the mine were dissatisfied with their treatment, which in turn cost Hinks money because of frequent strikes and a high rate of absenteeism.

By 1973, when a high-level UMWA safety specialist named Davitt McAteer proposed a truce, both sides were willing to listen. John and his fellow local union officers were called to a meeting with company officials and several consultants from Pennsylvania universities. The men sat all day in the conference room at a local motel, cigarette smoke and academic words thickening the air. The consultants talked about the movement in this and other countries toward ''autonomous work'' programs, the coal mine as a ''socio-technical system,'' the need to consider ''motivational factors.''

Finally, as suppertime approached, John cut into the presentation. ''What do you want?'' he asked. ''What exactly is it you want?''

Eventually, they worked out the terms of an experiment, based on the company's pledge to let the miners direct their own work. Each crew would figure out who should do which jobs on a given day, what supplies had to be ordered, when equipment maintenance should be done. The company would provide greatly expanded safety training, so that every miner would know the laws and the safe way to perform every job on his mine section. Each crew member would be paid top rate—a raise of up to seven dollars per day for the men with less seniority—and the company and miners would split any increased profits which resulted from the experiment. Instead of constantly fighting each other, the company and union would look for solutions to disputes that would satisfy both sides.

For the first time in the industry's history, coal miners would be free to put their own safety first. In turn, the company hoped, the

Rushton workers would be more skilled and happier with their jobs, and therefore more productive.

John couldn't believe his ears.

"The coal mines had been run the same way for a hundred years, and now they were talking about giving the men some control," he recalls. "I thought, if my dad, or my grandpap, could see this, they'd think it was a dream."

The right to direct their own work was initially given to the crews on only one of the mine's four sections. Then the program was extended to a second section. The men planned their work, switched jobs when someone didn't feel well, and helped each other with the messiest, most physical tasks.

John, who as a mechanic helped maintain equipment on the experimental sections, found that the program gave him a new attitude about his job.

"Instead of having one man trying to do the thinking for everybody, we had the whole crew figuring out what had to be done," he recalls. "It was exciting to see how it was going to work out.

"Before, the only safety training we ever had was what you learned from experience," he says. "That's how we got a lot of people hurt, because in a coal mine you make a mistake, buddy, and you may not be around to learn from it. Just by having six weeks of training in this program, I'll bet you these guys knew more about safety than any other miners in the country.

"The young guys were in favor of it, naturally, because now they were getting paid the same as everybody else," he adds. "And the older guys, they had never seen anything like it. It's like one guy said: it wasn't just work anymore. You actually enjoyed mining coal."

During the first year of the experiment, violations of federal safety laws were cut in half, the injury rate was well below the national average, costs for production and maintenance dropped steadily, the rate of absenteeism fell to about one-fifth the normal rate for the industry, and the mine was virtually free of strikes. In addition, seven thousand extra tons were produced as a result of the program, according to calculations by the company and the union.

The program was eventually expanded to all four of the mine's sections. As a result of a local union vote, however, participation was made voluntary for the miners. Any miner could choose to work at the old, lower pay rates, do only his assigned job, and be told what to do. As might be expected, few miners chose this option.

The Rushton experiment was working, but only as long as both sides lived up to it.

"Blair Rickard [the mine superintendent] and the company's section bosses never wanted to carry out the experiment after Hinks set it up," John says. "They didn't want to give up their authority. See, they were afraid they might be out of a job someday if we could direct our own work. Plus they knew they were going to be judged on how much coal was produced, whether the mine was safe or not.

"Hinks rarely comes to the mine, because really all he cares about is how much money they've made at the end of the month. So the bosses, they were free to start ignoring the experiment unless it was to their advantage. Sure, you could direct your own work when everything was going right. But when you'd have a machine in bad shape or you'd hit

bad conditions, instead of letting you take the time to work safely, like they promised, they'd be pushing you just like they always did.''

The company began to use the special classroom training sessions, which were supposed to focus on safety, to talk almost exclusively about ways to increase production. When job switching within a crew would mean assigning a better operator to a machine, the foremen would allow it. But when the miners wanted a switch to promote safety—to teach a miner about the hazards of another job or to give a machine operator a break—the bosses would say, ''Not right now. We haven't loaded enough coal today.''

The mine's injury rate began to rise sharply. Perhaps because the productivity rate was still excellent, however, Hinks made little effort to force his foremen to abide by the rules of the experiment he had helped establish.

When Hinks admitted at one meeting that most of the foremen frequently disregarded the experiment, John asked him, ''If the bosses don't do what you want, why don't you send them down the road?''

''You have to remember that there's only so much I can do,'' Hinks replied. ''After all, foremen are children of God too.''

With the experiment's original spirit of cooperation gone, the local union and the company reverted to confrontations and power struggles to resolve their disputes. The first major clash occurred in early 1976 when an older worker asked to be shifted to a vacant job above ground. Company officials tried to award the position to a less senior man who had learned it while trading jobs as part of the experimental program. The local union filed a grievance, but the company refused to follow the seniority provisions of the UMWA contract.

After a strike began over the issue, the company tried to fire ten miners, and later tried to force the local to remove John and another man as local officers.

''Their muscle against our muscle,'' John recalls. ''It was still the same old game.''

The company eventually lost that round, but it won the next one. The miners became incensed when a boss broke an unwritten coal industry rule by pushing a worker during an argument. The next day, union leaders met with Superintendent Rickard to discuss the incident. As they talked, a foreman came in to report that a strike was about to start. John and the other local officers went to the mine entrance, and asked the miners congregated there to go in to work while the two sides finished meeting. But the miners wouldn't listen.

''You're not going to get anything by meeting,'' said one older miner. ''If we work, they'll think we're going to let them get away with this. Let them make the first move. Let them suspend their boss. Then we'll work.''

Some of the young men had already headed for the parking lot, where they were making plans to go fishing. Any excuse for a few days away from the mine was good enough for them.

When the officers told Rickard the men were on strike, he refused to continue the meeting, invoking the usual coal company policy: ''no work, no talk.'' The miners, in turn, would not work until the meeting was completed. As often happens in the coal industry, what could have been a one-day strike stretched out over two weeks.

The miners gave in first by returning to work, and the meeting on the original grievance was reconvened. Rickard asked everyone to bow

their heads to pray that "this meeting will be concluded in the light of God's wisdom." When the union asked that the boss be suspended from supervisory duties, the company offered only to promise that he would not commit the same offense again.

"Each one of our guys had lost about six hundred dollars during this strike," John recalls. "We were trying to get something to show for it. But the company had us, and they stuck it to us all the way."

As the local union officers left the meeting empty-handed, Rickard tried to control his smile. "Well," he said, "I guess you can go back and tell your men that about the only thing their union does for them is collect their dues."

A few weeks later, in April, 1977, John helped convince his local union not to send its officers to any more meetings about work experiments at Rushton. The miners wanted to embarrass Hinks, who had received a great deal of favorable publicity because of the experiment, but they didn't want to end the program altogether. Even if the miners only got the chance to direct their own work part of the time, they had

BUCK KOPTCHAK

BERNICE DOMBROSKI

more freedom than under the old system. If nothing else, the program continued to mean higher pay for most of the younger workers. The experiment wasn't really working out, but, like some unhappy marriages, it was kept alive because there was no better alternative.

Sitting at the edge of the "dinner hole" where a crew of miners is eating lunch, Mike Repaski gives a visitor to the mine his standard speech on the Rushton experiment. As the company's training director for the mine, Repaski has traveled with Warren Hinks to more than a dozen conferences all over the United States to boast of the enlightened philosophy of Rushton's management. By now, the phrases roll right off his tongue.

"The coal miner of today is a different breed from fifty years ago, and we've had to change too," he says. "You can't run a dictatorship in the mines anymore. You have to reason with people, let them get involved.

"You don't need a hammer over their heads," he says. "That's the great thing about our program—we're not trying to use that hammer anymore."

As Repaski talks, the miners lounge on the dusty mine floor or sit on timbers or mine machinery, eating their sandwiches and candy bars with grimy fingers. The only difference between the dinner hole and other areas underground is that special care has been taken to support the roof over it. There are no tables and chairs and no washing facilities.

During lunch, John talks with Bernice Dombroski, the newest member of the crew. Bernice and her husband, who became disabled with a back injury many years ago, used to live near the Socoskis in Osceola Mills. One of more than a thousand women who have entered the coal mines since 1973, and the first woman miner at Rushton, she has proven to the men that she can do the work.

"I'm thinking about trying for a job on one of the machines now," she tells John. "I could use the extra money."

"Why not?" John encourages her. "You do that. You sign up next time there's an opening."

When lunch is over, the mincrs head back to work. With no water on the mine floor, solid roof over their heads, and no production equipment broken down, this crew is allowed to go about the job without much supervision. The experimental program is still in evidence here, yielding a relaxed atmosphere unusual for a coal mine.

John goes to figure out why the controls are sticking on one of the "jeeps," the flat vehicles used to transport workers throughout the mine. Bernice carries and sets timbers for temporary roof support, and spreads rock dust from twenty-five-pound sacks to keep down the coal dust, while Buck Koptchak helps insert steel rods into the roof to bind the layers of rock together. Then it is safe for the continuous miner operator to enter the area.

John says the wild-looking young man who is the miner operator on this crew was once fired after he was arrested at his home for allegedly dealing drugs. The union officers got his job back, arguing that he could not be discharged for conduct outside the mine.

"A lot of mines have a big drug problem, and we don't want one here," John says. "Hey, you know how I'd feel if somebody was selling drugs to one of my kids. But there was a principle involved. If he could be fired for what he did at home, so could any one of us."

JOHN SOCOSKI, JERRY MORIARTY

John gets the jeep fixed, and is told to take it through the network of narrow tunnels to another mine section. After a bouncy, ten-minute ride, he comes to the end of the track. Grabbing his tool box, he steps off the jeep into a deep puddle. He walks toward the continuous miner, his body bent because of the low roof, his boots slipping on the soggy coal.

''How do you like our water wonderland?'' calls Jerry Moriarty, the miner operator. ''Got your fishing pole?''

Jerry turns over the controls of his machine to his helper, and sits down to talk. On this section, there is tension between the miners and bosses as though there never had been an experiment at Rushton. Up to seven million gallons of water pour from the coal seam each day, making it difficult to maneuver the machinery and increasing the chance of electrocution. The miners feel they are forced to run their machines when the area should be pumped out first. But Superintendent Rickard doesn't see why the wet conditions should be allowed to slow down production. ''Bad conditions,'' he likes to say, ''is not a reason. It's an excuse.''

''You can have all the experiments you want,'' Jerry says, ''but if you really want to do something about safety, you'd have to fire the superintendent and the bosses. Then you could start a safety program.''

He and John laugh at the thought, but he is serious.

''There's nothing the bosses do that we can't learn to do,'' Jerry continues. ''We could elect our management, and vote them out if they weren't doing a competent job.

''You know, that could be a good idea for the UMWA,'' he says. ''Go to the biggest, roughest non-union areas, and start our own mines. You could pay the miners more and run the safest mines in the country. I know you could; we did it for a while here when the experiment first started.''

John smiles and shakes his head from side to side.

''No, Jerry, it'll never be tried,'' he says. ''Anything new, somebody's going to make an issue of it, because all we know is what we've always done.''

Both John and Jerry have heard of the better safety records in countries with nationalized mines, like Great Britain, and in socialist countries in eastern Europe, but they don't know anything about how the mines are actually run in those countries. Neither of them has ever traveled abroad to see for himself, and the UMWA has never made any real effort to learn from unions in other nations.

"Socialism—that's what I said our program was when I first heard about it," John says. "Sharing the authority, paying everybody the same thing, that's the beginning of socialism. We were able to try something different because we're just a small mine and nobody paid that much attention. But on a larger scale, it would never get off the ground."

Later, outside the mine, Mike Repaski, the training director, picks up where he left off at lunchtime, talking about how "at Rushton we've never been afraid to give the men a voice." In an attempt to short-circuit his monologue, he is asked how he can work for a company like this that promotes *communism*—giving the workers control over the mine. Suddenly, his cheerful discourse on modern labor-management relations is over; he does not see that he is being deliberately provoked.

"Now just a minute," he snaps, his face tightening. "We didn't give them any control over the mine whatsoever. We just told them they could go ahead and do their jobs if they knew what to do, which is the way it's always been.

"The only power we delegated to the men in this program," he says, "was the power they already had anyway."

Before I'd Be A Slave

After being shut inside all winter around their coal stove, the Martin family was always ready for the Emancipation Day picnic. Each year on April 9, the schools for black children were let out for the day, and black miners were excused from work. From Number 1 camp, Number 2, and Number 3, all the way up to Number 14—the clusters of company-owned houses named after the U.S. Steel mines nearest them—hundreds of black families like the Martins walked to the ball field for the party.

The weather in southern West Virginia was usually perfect at this time of year, warm but not as humid as it would be in a month or two. The kids crowded around the ice cream tubs, clamoring for what was a rare treat in these Depression years. The grownups took advantage of U.S. Steel's generosity by going back for frequent refills of free ice-cold Fort Pitt beer, and cheered and hooted as their favorite local boxers and wrestlers demonstrated their skills in exhibition matches.

When the crowd had been properly plied with alcohol, the speeches began.

"You all know that April 9 was Emancipation Day, the day Lincoln freed our people from slavery. And I think today is a good time for us to remember how much U.S. Steel has done for our people too."

The speaker went on to remind his listeners how much better off they were here in Gary than they had been before U.S. Steel brought them in boxcars from Alabama or Georgia. There, they had to work in the fields for almost nothing; here, the company provided them with a house to rent, a doctor, a store, and a job—as long as they followed the rules.

The next subject was politics. The company didn't want any employees who voted for Democrats, the audience was told. Everyone ought to cooperate with the company's poll watchers to see that the Grand Old Party of Herbert Hoover and Alf Landon got unanimous backing from the good folks of Gary.

Saving the best for last, the company then brought on the top officials of the "independent" union they had established for the Gary mines. Emancipation Day, the officials said, was a fitting occasion to resolve once again to keep the United Mine Workers out of Gary. We must help the company keep UMWA organizers off our roads and out of our homes; otherwise, how can we expect to continue to have such fine celebrations as this?

Young Napoleon Martin, Jr., knew his father and uncle were secretly members of the UMWA and didn't care for these speeches. But he looked forward to the food and entertainment each year, and was sorry when the tradition ended. In December, 1941, an organizing drive led by UMWA President John L. Lewis forced all the major steel

N&W

companies to sign UMWA contracts for their mines. After that, just as the speakers had warned, U.S. Steel no longer was interested in sponsoring an Emancipation Day celebration in Gary.

"You know, I never will forget it," recalls Napoleon, now forty-nine, as he leans back in an easy chair in the living room of his old company-built house in Gary. "At the end of the day, when they had us all in the right mood, they had us sing this song: 'Oh freedom, oh freedom. Before I'd be a slave, I'd be buried in my grave.' Everybody sang it out real pretty.

"Yeah, oh freedom, oh freedom. And here it is, thirty years later, and you can see for yourself we're still fighting for it."

Gary, West Virginia, was a home-made coal town, built according to U.S. Steel's special recipe. When company officials decided at the turn of the century that they needed Gary's high-quality coal to help fuel their steel mills, they knew that a key ingredient for a successful operation was an obedient work force. So in addition to hiring European immigrants, they sent representatives to the Deep South to find black farm workers, who knew a great deal about hard work and nothing about unions, and who could not expect to find jobs elsewhere if they offended the company.

The company built small wooden houses that resembled military barracks for their miners, intended not so much as homes but as temporary shelters while the workers were off-duty. As long as a miner was working, he could rent one of these houses, but if he became sick or disabled, or was suspected of being a UMWA supporter, his family could be evicted, with no alternative but to leave the area. Armed guards called "shack rousters" came through the camps in the early morning hours to check on those who had not shown up for work. A miner might not be feeling well, but if he wasn't on his deathbed, the shack rousters would force him to go to the mines.

The miners were paid in "scrip" that could only be used at the company stores, which provided everything from food to clothing to furniture. Prices were set so high that the company made back in store profits much of what it paid out in wages, and most miners were constantly in debt to the company.

U.S. Steel was the only employer in town. The doctor worked for the company, and was paid by automatic deductions from the miners' paychecks. The barber, the baker, and the schoolteachers all were company employees. There was no town government other than the U.S. Steel superintendent, who even became involved occasionally in resolving marital squabbles or other disputes. When people died, they were buried in the company graveyards.

Napoleon's father, Napoleon Bonaparte Martin, Sr., and uncle, Willis, came to Gary in 1926 when hard times hit the western Kentucky coalfields where they had been employed in UMWA mines. If they hadn't been in such need of work, they wouldn't have stayed in Gary after seeing what conditions could be without a real union. U.S. Steel had set up a company union, with hand-picked officers; union "elections" were conducted by standing vote, and those who didn't stand for the incumbents risked losing their jobs. Company guards patrolled the camps looking for UMWA organizers, and at times state policemen camped in tents in Gary to help keep out UMWA supporters from other parts of McDowell County.

"It was just like what you hear about concentration camps under Hitler," recalls seventy-eight-year-old Willis Martin. "A lot of the people were always running to the company to tell on their neighbor, hoping it would save their own skin. If you said hello to a United Mine Worker on the street somewhere, the next day when you went to work you had a message to go see the superintendent. He'd ask you, 'Who was that you were talking to yesterday? What did he want? What did you tell him?'

"People didn't know what the Mine Workers could do for them, and they turned against it because they didn't want no trouble," he says. "The company paid them a few cents more a ton than UMWA, but what the men didn't realize is that they were working thirteen or fourteen hours a day, six days a week, doing all the timbering and laying track and cleaning up for free, while the United Mine Workers was just loading coal and going home after eight hours. So the company was actually saving an awful lot of money this way.

"We used to go to work so early in the morning and come home so late that on Sunday morning you'd see a little baby start crying when he saw the strange man in the house. 'Mama, who is that man?' that baby would say, because it hadn't seen its own father for so long."

U.S. Steel arranged recreation for the black community, hiring semi-pro baseball players to mine coal in the winter and entertain their neighbors on the ball diamond in the summer. Old-timers can remember when Satchel Paige and Josh Gibson, stars of the professional Negro Leagues, brought their barnstorming tour through Gary. The company also allowed its favorites in each camp to run poolrooms and bars, knowing that they could be trusted to report their customers' conversations.

Special privileges were reserved for white workers. They had better restaurants, bowling alleys, and tennis courts. Several camps had movie theaters, with separate sections for whites and blacks; if the theater was crowded, whites could use the black section, but blacks could never cross into white territory.

"The poor white kids were supposed to stay happy because they had a little more than we did,'' says Napoleon. ''But you know, they got along better with us than they did with the bosses' kids.

''We used to play football with them. They had their Boy Scout team, you know, with regular football equipment. We played them in our old raggedy overalls, and we always beat them. Their coach called it 'practice games' for the real league they were in.

''Another thing we did was take some old grass sacks and fill them with dirt and stop up one of these creeks. We'd make us a swimming hole, black and white together. It was the only way to stay cool in the summertime. But, you know, as fast as we'd build our little dam, the company always sent some guards to tear it down. I never did know why.''

After graduating from the black high school, Napoleon went to work in one of U.S. Steel's mines. In his years underground, he saw for himself how valuable the UMWA could be. Wages had been raised and hours shortened after the company signed with the UMWA in 1941. Several long national strikes led to the creation of a fund to help cover miners' medical costs and pay the first pensions in the coal industry. The miners had also won the right to elect a local union safety committee and a ''checkweighman'' to keep the company from cheating when it recorded the tonnage mined by each worker.

But after Napoleon was drafted for the Korean War in 1951, his eyes were opened to the need for even more progress at home.

''We had integrated units in the Army,'' he recalls, ''and we got along fine on the base during training. Of course, when we went to town, we didn't have nothing to do with each other. We were stationed mainly on Okinawa, anti-aircraft, protecting our planes making raids. I guess we've learned since then that we can't run all around the world telling people what kind of government to have, and I still blame that war and Vietnam for a lot of the inflation we've had. At the time, though, I believed in what we were doing.

''But then we came home,'' he says. ''We'd been fighting side by side for two years, so some of the blacks and some whites went into a bar in California together, and you should have seen the trouble that started. I was sick of it, having to sneak around to the back of some restaurant to have them hand me a sandwich.''

Back in Gary, Napoleon decided to take some action. With a group of friends, he went to the drugstore, which had one soda fountain for blacks and one for whites. They walked in and quietly sat down in the white section.

It was 1953, two years before Rosa Parks staged her famous sit-in on a bus in Montgomery, Alabama. The drugstore manager was baffled: why would these fine young men he had known all their lives do something crazy like this?

''Well, it was more or less of a stalemate,'' Napoleon says. ''They wouldn't serve us and we wouldn't move. But there wasn't no incident or nothing, not in a little place like Gary. So what they did, they just closed down the white section of it, and finally they closed down the store altogether.''

Soon after the war, the U.S. Supreme Court ruled that ''separate but equal'' school systems violated the Constitution, and a few blacks broke the color line at Gary's white high school. But discrimination continued in another important area: treatment on the job.

U.S. Steel, like other coal companies, was mechanizing its mines, introducing automatic cutters, loaders, and conveyor belts. At the same time, electric utilities, factories, and homes began using oil instead of coal, and the recessions of the late 1950s reduced demand for coal and steel. The UMWA work force fell from four hundred fifty thousand in 1947 to one hundred seventy thousand in 1960.

Blacks generally had not been given a chance to move into jobs as machine operators, so they bore the brunt of the layoffs. Many relatives of Napoleon's wife, Leontine, had to move to Philadelphia and Wilmington, Delaware. A few housewives even went temporarily to New York City to work as maids, and some miners who lacked only a year or two to qualify for pensions were forced to drop everything and move north.

Napoleon and his brother, Francis, should have had little chance of being laid off, since both of them had jobs at U.S. Steel's Gary preparation plant, where coal was processed for shipment. But they had become very active in their local union, filing grievances and challenging discrimination, and had irritated both the company and the UMWA district.

In 1959, Francis was fired on charges that he was not doing his job properly. After union officials tried unsuccessfully to exclude him from his discharge hearing, he forced them to take his case to an arbitrator, who ordered him reinstated.

Then word was sent to Napoleon that if he didn't stop causing problems at the plant, he would be laid off. The layoff came, and he had to threaten to file charges with the National Labor Relations Board against both the company and union before he was given his job back.

''By trying to enforce the contract, we weren't just making things hard on the bosses,'' Napoleon explains, ''We were giving the district a lot of cases to handle, and it was making them look bad if they compromised or lost a case where we were right.

''All the district people in the union were appointed at that time—nobody was elected. And the top leadership, they didn't want intelligence. They wanted to appoint people that weren't ever going to be after their job.

''John L. Lewis, he was a respected man, but as far as black people speaking out, we didn't fit into his scheme of things. They only had one black working for the district, and he probably had a sixth-grade education.''

Without union backing, it took Napoleon until 1968 to become the first black to work in the cleaning plant's electrical department.

''They think blacks have a natural ability for certain types of jobs, like with a shovel in your hands or the janitor in the office, but then the whites make better operators for your sophisticated control panels,'' he says. ''One time when I was classified as a laborer, I tried for a better job and they gave it to a white boy with less experience and education than me. They said he was 'too valuable' to be just a laborer.

''Don't get me wrong, I knew how important the union was,'' he adds. ''But if you talk about the union responding to our needs, it wasn't.''

When John F. Kennedy stopped in Gary while campaigning in the 1960 West Virginia primary, he made a speech about what he had been seeing in the state. He talked about miners sitting on the porches of dilapidated houses, with no work and no benefits, and he talked about the scarcity of good health care and the lack of transportation.

Local coal company officials, politicians, and newspaper editors were furious: how dare he talk about our proud state that way! The miners, on the other hand, loved it. If they had any complaint, it was that he hadn't assigned the blame for their condition on the coal companies which had laid them off without making any provision for them, and which had always discouraged the development of other industry in the region.

When Kennedy declared a War on Poverty, U.S. Steel responded, in its own way, to the patriotic call. As mining families moved out of state to find jobs, company officials solved the problem of substandard housing by tearing down thousands of homes, along with most of their company stores and other businesses. Their campaign of destruction helped them save on maintenance costs and make way for expansion of the company's strip mining operations. Stripping, which involves bulldozing enough of a mountainside to expose a coal seam and then trucking the coal away, is the most profitable way to mine because it requires fewer workers.

STRIP MINING, BOLT MOUNTAIN, WEST VIRGINIA

"Stripping cut down on jobs even more," says Napoleon, "and now the people who live around here are stuck with landslides and mud and run-off into the creeks. Even if you follow tight regulations, you can't strip these steep hills here in West Virginia and not have landslides, and in those days, there weren't even any regulations."

The removal of so many houses came back to haunt the region when, in the late 1960s, increased demand for coal and steel began to accomplish what the poverty war never had. As the coal industry expanded, thousands of miners who had left the state could find jobs at home again, but there was nowhere for them to live. Rather than tackle the new housing problem, U.S. Steel officials decided to cut costs further by selling the houses they still owned to those who were living in them. Then they created the City of Gary, a taxpayer-financed town, to take over operation of the Gary electric power plant, water system, and sewage plant, all of which had become increasingly costly to operate. They also gave twenty thousand dollars to start the city off on the right foot—barely enough to pay half the cost of a new fire truck.

If the housing shortage were going to be solved, then, the mining families themselves had to take the initiative. Many were living in mobile homes which were relatively cheap, could be installed quickly, and required little ground, but which were also firetraps, unstable in wind, and too small to provide much space for furniture or peace of mind. Many young families who couldn't or wouldn't buy trailers lived in equally tight quarters with their parents. Napoleon and Leontine were forced to stay in their rundown company house because there was no alternative.

The few new houses that were being built were too expensive for miners. Nationally, the cost of a new home had been rising about twice as fast as the average family's income, so that only twenty-seven percent of families could afford a median-priced new home. If anything, prices were rising even faster in McDowell County.

A small group of Gary residents, including Napoleon and Francis Martin, began meeting with Deborah Tuck, a UMWA housing specialist. They decided to set up a non-profit housing corporation, with representatives from UMWA local unions and the community on the board of directors. Their goal was to build single-family homes that miners could afford. No one would make a profit on the sales, and the development would be completely integrated—black and white, young and old, lower- and middle-income.

At meeting after meeting, the same handful of men and women wrestled with the details, calling themselves the Gary Bicentennial Building Association. They learned how to select a site for homes, to pick a builder, to hunt for state and federal funds. They faced skepticism from many—no group of mining families had ever undertaken a similar housing project—and opposition from some because "all you're doing is trying to build houses for some niggers."

But the major obstacle they faced was obtaining land. In West Virginia as a whole, two-thirds of the privately-held land is owned or controlled by out-of-state companies. In McDowell County, six corporations own seventy-one percent of the land. Pocahontas Land Company, a subsidiary of Norfolk and Western Railway, owns more than one hundred thousand acres, or nearly one-third of the county. In Gary, Pocahontas and U.S. Steel own virtually every acre, and the two work together hand-in-glove; one has the land to lease for mining, the other has the ability and need to mine it.

When the housing group asked U.S. Steel for help in obtaining land, company officials said they couldn't influence Pocahontas. When

the group went to Pocahontas, asking to buy a particular thirteen-acre site, company officials drew up a contract which provided that they would sell the surface rights but retain the right to lease the minerals under the land to U.S. Steel for mining. The residents would have to promise not to seek compensation for damages caused by noise, air or water pollution, dust, or vibration from mining under or near the land.

''I think they hoped we would turn down their contract,'' Napoleon says. ''They were afraid if they sold us this land, someday they'd have to sell us more for more houses. They prefer to keep it for mining, or in case they decide to have some rip-off builder come in here and put up homes for eighteen thousand dollars they could sell to the miners for forty-five thousand.''

The association eventually had to agree to the contract, and in November, 1978, the first of twenty-six homes was ready for occupancy. They would have an average sale price of less than forty thousand dollars.

Napoleon and Leontine bought one of the first houses. At last, they would have the kind of housing many Americans outside the coalfields take for granted.

''We worked a long time for this,'' says Napoleon. ''The whole community—everybody worked a long time for this.''

But the group had only begun to help ease the housing shortage in McDowell County. At least ten thousand of the county's fourteen thousand homes were built more than forty years ago. By the early 1980s, the mining work force is expected to increase by more than two thousand. An estimated six thousand new homes will be needed in the county by that time.

The housing association meetings are usually held right after the meetings of the Gary health association because many of the same people are involved in both groups. Until April, 1976, the Gary area had no medical clinic or hospital, and the aging company doctor could not handle many patients. The only available facilities were a hospital and clinic in Welch, the county seat. Both had long lines and limited staff.

Again, local union representatives, retired miners and miners' wives created a non-profit corporation, this time with the aid of Craig Robinson and Tom Ludwig, UMWA health experts. Leonard Callahan, a retired black miner who had been president of his local union for thirty-three years, was elected health association chairman. The group had to go through a process similar to the housing association's: applying for government and foundation funds, finding a building that wasn't being used, arranging for a doctor through the National Health Service Corps. With the support of Napoleon and other board members, Martha Chapman, a thirty-one-year-old white woman who was working for the UMWA Health and Retirement Funds, was chosen clinic administrator. Her husband, Claude, is chairman of the local union health and safety committee at the Gary preparation plant. The other employees, including the nurses, also came from miners' families.

In less than three years, the clinic had two doctors (one black, one white), and a dentist. Patients learned that the clinic could give more personal attention than the larger medical facilities in Welch. Each person was given a complete physical examination on his or her first visit, leading to the discovery of many developing health problems that could be treated before they became more serious. Past prescriptions

NAPOLEON AND LEONTINE MARTIN WITH NEW HOUSE

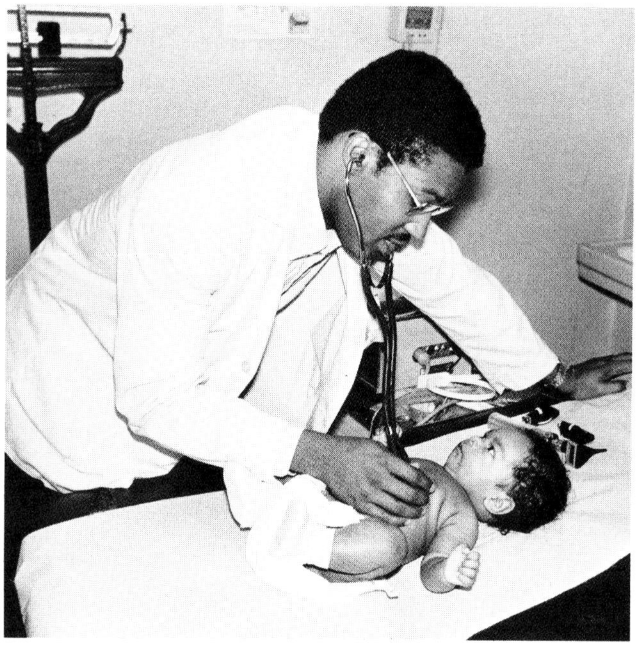

written by other doctors were reviewed to make sure the drugs being used were really necessary and effective.

"The community that uses the doctor hires the doctor," says Napoleon, "so we know we're going to get good service. It used to be that when I was too sick to work, I'd have to sit and wait all day over at Welch, or just forget about seeing a doctor. At our clinic, they set up an appointment schedule, so I know when they'll be able to see me."

The clinic houses a health education project, funded by the federal government with participation from all medical facilities in McDowell County. The goal is to teach people how to maintain they own health by improving their diet and by setting aside enough time for exercise, sleep, and relaxation.

"In some ways, people used to live better than they do now, even with all our modern conveniences," says Martha Chapman. "My mother is seventy-seven, and she's helped raise four generations after her. We always lived on a farm—I was the last of fourteen kids—and we never had a whole lot of money, especially once my dad had black lung and couldn't work.

"But even though we may not have had anything fancy, what we had to eat was a lot better for us than the food we eat now," she says. "And farm life was a healthier atmosphere. You did your work, but you weren't wearing out your nerves chasing that big money the way everybody is today."

The clinic association approached Pocahontas Land Company for a vacant piece of property where a larger clinic could be built with an adequate parking area and a pharmacy. Pocahontas wanted the association to agree in writing, in advance, to pay whatever sale price an independent appraiser set for the property, whether it be two thousand dollars or two hundred thousand dollars. As expected, the association turned that down.

If land for a new clinic is every obtained, Martha hopes to use the current clinic building as a childcare center. In Gary, as in the rest of the country, more and more women are working outside the home, yet the problem of how to provide proper care for their children remains largely unsolved. Some kids stay with grandmothers, older children, or hired babysitters; others spend several hours after school each day unattended.

"You can't dictate to the women anymore," Napoleon says, admitting that he wasn't thrilled at the prospect of helping more around the house when his wife took a job at Stevens Clinic near Welch. "I can see why she wanted to get a job. She was tired of being around, talking on the phone, making pillows and these afghans. She wanted to get out."

Claude Chapman said the fact that both he and Martha work doesn't affect their daughter, Paula. He gets off work at three and can be home by the time she arrives from school, and both grandmothers live within a hundred yards.

"I couldn't expect Martha to spend every day cleaning this damn trailer; anybody'd go crazy doing that," he says. "The funny thing is, a lot of guys who say they don't want their wives to work are the same guys who go off every weekend fishing or drinking. We don't do that. We both work, but then when we have free time, we spend it together, going to a fiddlers' convention or something like that."

Working women are becoming accepted by most people in Gary, as long as they stay in traditional female careers. But when women started getting jobs in the mines and at the preparation plant in the past several years, there was open warfare.

Not long after a married woman in her early twenties became the first to be hired at the plant, Martha started getting phone calls at work from miners' wives. The callers had heard a rumor that Claude had taken the woman to the plant's dump during one shift and that they had been gone, by themselves, for three hours.

"We knew this would happen," the callers said. "We knew we shouldn't let U.S. Steel hire these women to work with our men. If you're a real woman, you'll put a stop to this. We want to know what you're going to do."

"The wives were threatened and the men were having a good time keeping them up in arms," Martha recalls. "Everywhere I went, I'd run into guys from the plant who wanted to add their own details to the story."

The rumors continued to circulate until the woman's husband beat her up so seriously that she had to miss four days' work. But she came back to her job, black eyes and all, and when four more women were hired at the plant, not too much was said about it.

MARTHA CHAPMAN AND HER MOTHER, OLLIE BLEVINS

There's not much to do in Gary. In the afternoons, twelve-year-olds can be seen sitting in small groups, smoking marijuana. On Saturday night, Napoleon and some of the older miners talk in front of the old company store, while across the street, teenagers play basketball with an imaginary hoop. In this town with a population of three thousand in the city limits alone, there is no theater, restaurant, dance hall, pool hall, bowling alley, or any other place for young or old to get together. The only businesses are two food stores, a state-run liquor store, a beauty shop, a service station, and a tire store. People who want to start other businesses find that Pocahontas Land Company and U.S. Steel won't sell them any land. Many residents believe that U.S. Steel won't allow any other employers in town because it doesn't want competition for labor.

George Deskins, a retired coal miner who was the town's police chief until he died in the spring of 1977, said that Gary was ''just lucky'' not to have more crime than it did.

''You can't tell people to get off the streets when there's no place else for them to go,'' he said. ''We have some marijuana among all ages—I know they have it at the cleaning plant and in the mines too—but between you and me, I don't worry about that so much, because people are actually more easygoing with that than what they are with booze.

''But the thing that scares me is hard drugs,'' he said. ''A lot of other communities around here have a problem with that, and I just wonder how long until we're going to have it too.''

Charles Hodge, the mayor of Gary, wants the city to build recreational facilities, but that will take money as well as land. Since the city council was created in 1971, it has not had much leverage on the only source of real wealth in town: U.S. Steel. Hodge is the only United Mine Worker on the council, although the great majority of town residents are coal miners. Until the elections in June, 1977, more than half of the council members were supervisory personnel for U.S. Steel, who were unlikely to support any demands on the company for contributions to the town. At one council meeting, these members even refused to have their picture taken with the other representatives, saying that ''the [U.S. Steel] public relations people told us not to get involved in any publicity without checking with them first.''

Mayor Hodge has been trying to get money for recreation from the state or county, but low taxation of the coal companies has left them relatively poor as well. A 1974 study by the Huntington (W.Va.) Herald-Dispatch found that West Virginia loses at least one hundred fifty million dollars each year because of inadequate taxation of coal and other property. The study found, for instance, that the McDowell County assessor's office had increased the appraisal of a twenty-one thousand acre tract owned by Pocahontas Land Company by only four dollars and fifty cents per acre in the previous thirty-seven years. The appraisal on a seven thousand acre tract owned by Consolidation Coal Company had not been changed at all in fifteen years.

''In this state, the companies control the tax officials and the politicians, and the politicians control the voters, instead of the other

GARY MAYOR CHARLES HODGE

way around,'' says Napoleon. ''I know because I used to be in it myself. Just before election time, they send in a lot of money, and it buys them the votes they need.

''If you're a poll watcher, what you do is you 'help' the people vote. You pull the levers for them, and you make sure your candidate gets the vote. The people don't know the difference.

''I got out of it when I realized that it wasn't helping the people around here,'' he says. ''You think if you help this particular politician, it's going to help your area, but it doesn't, because he doesn't owe you a single thing when election day is over. You've got him for four years, and the only people he owes anything to are the people who gave him the money to start with.''

Napoleon says that the companies use their political power not only to keep their own tax assessments low, but also to see that tax money collected from the public is spent to benefit them.

"Look at all the millions of dollars that are being spent now on vocational training programs in the schools; that's the latest thing in education," he says. "The company gets its training done free. They send their equipment to the schools to be repaired, and their bosses make extra money as instructors.

"The kids, they can't read or count—they're getting a worse education than when I was young. But U.S. Steel, it knows it's got its next generation of coal miners. That's the kind of thing that ties up all the money so there's nothing for things like recreation."

Because of the lack of recreational facilities, most people in Gary turn to their TV sets. In many families, the TV is on virtually every minute that someone is home, even if no one is watching it. There generally seems to be little interest in most of the programs; often, people who are sitting in front of the TV cannot tell a new arrival what the current show is about.

An exception was the series, "Roots," which, by describing one black family's past, traced the history of American race relations during the era of slavery. Each day at the mines and preparation plant, there was a great deal of discussion about the segment shown the night before. Many blacks felt the series brought out into the open both past events and black-white feelings which continue today. But Napoleon, who had helped lead Gary's first sit-in twenty years before, didn't watch a single episode.

"My son had given me the book, and I could tell what Hollywood was going to do with it," he explains. "I knew they would pick out the parts that would get all this old hatred stirred up, and that's exactly what they did.

"Down at the cleaning plant, the young blacks and the young whites were arguing, wanting to fight each other because of something that happened two hundred years ago. Only the younger people—the older guys don't want to fight about that stuff anymore."

Napoleon watches less TV than most people in Gary because he has another hobby: the union. As a grievance committeeman since 1963 and now as local union president, he spends many of his evening and weekend hours visiting other workers and discussing problems at the plant with the company.

Napoleon also devotes a large amount of time to UMWA politics. Although he suffered from the policies of John L. Lewis and his successor, W. A. "Tony" Boyle, particularly in their refusal to allow elections for district officials and their treatment of blacks, he actively supported Boyle against insurgent Jock Yablonski in the 1969 UMWA presidential election, helping to round up a big turnout for Boyle in the Gary area.

"The leadership was in control of the union, and I needed the support of the union against the company," he says. "Sometimes you might have to go along with somebody when they need you, even if you don't agree with everything they're doing, because then when you need them they're going to help you."

Yablonski lost, but vowed to tell the courts all about the union corruption he had seen as a UMWA executive board member. Boyle ordered him killed, and three weeks after the election, Yablonski, his

CLAUDE CHAPMAN, NAPOLEON MARTIN

wife, and daughter were murdered in their beds by hired gunmen paid with union funds.

After a three-year court battle, a federal judge found that Boyle had illegally used the union treasury and staff to influence the 1969 balloting, and a new election was ordered. A reform group called Miners for Democracy, or MFD, nominated rank-and-file miners Arnold Miller, Mike Trbovich, and Harry Patrick to run for the union's top three jobs.

In addition to the issues of violence and misuse of funds, the 1972 election involved other basic principles. The reformers said the union should give its members the right to vote for district officials and on new contracts, educate them about their safety and contract rights, involve them in legislative lobbying on broad social issues, and open the union magazine to all shades of rank-and-file opinion. The rank and file might make mistakes, but in the long run only they could keep their union officials on the right road.

Many Boyle supporters, on the other hand, believed that the union should retain its top-down structure. Only the union leaders were really in a position to know if a new contract was the best that could be negotiated, or if a certain miner was qualified for district office, or if a particular politician should be endorsed. Union members might be well-intentioned, but they would never be sophisticated enough to make those kind of judgments.

At least as importantly, patronage and power were at stake in the UMWA election, just as they are when the Republicans and Democrats compete for the U.S. presidency. From the international level to the districts and local unions, people who had worked hard to gain position or influence had a lot to lose if a group of unknowns suddenly by-passed the system and swept into office.

Napoleon again supported Boyle.

''I didn't want to take a chance with someone right out of the mines becoming the international president,'' he explains. ''I thought we were better off to stick with what we had.

''The union needs a boss. You can talk trash about dictatorship, but sometimes all you get with democracy is a lot of confusion.''

After Miller won with about fifty-five percent of the vote, virtually all of the democratic rights MFD had promised were established. The first elections were held in Napoleon's district, for example, and his brother, Francis, was elected as representative to the UMWA international executive board.

But the war between the factions never stopped, and reformers' attempts to improve basic union programs were gunned down in the crossfire. Non-union mines were springing up in the West, but an expanded organizing drive became bogged down in political debate over personalities and budgets. Coal companies were violating the UMWA contract with increasing regularity, and miners were increasingly turning to strikes in retaliation, but a program developed to educate local officers in contract enforcement without strikes was never adequately supported with funding and staff.

According to some reformers, the breakdown of union programs was entirely the fault of the Boyle faction, which undermined every initiative, regardless of its merits, in hopes of one day regaining control. Others charged that Miller shared much of the blame because he was unwilling to fight for reforms, gradually adopting Boyle's programs and style in an attempt to quiet the opposition.

FRANCIS MARTIN AT UMWA CONVENTION

But regardless of who is responsible for the union's troubles, the victims of the infighting have been the UMWA members and their families, as Napoleon and the people of Gary can testify. After first one side and then the other claimed that the union's multi-million-dollar treasury could not finance housing and clinic organizing, the specialists in those fields who had been brought onto the UMWA staff by the reform administration were forced to leave. They still provided advice to the Gary groups because they were concerned, but the union backing that made it possible for these projects to get started was gone, and the international union stopped sponsoring any similar projects in other communities.

In downtown Gary every day at lunchtime, U.S. Steel's white-collar employees congregate on a bridge across the little creek that runs through town. Above their heads is a sign with U.S. Steel's initials; a slogan, "Safety The First Consideration"; and a number indicating how many days it has been since the last injury at the company's Gary mines and preparation plant. The number of days shown on the sign often climbs as high as seventy or eighty, although reports by the company to the federal government indicate that there is an average of one injury every day at these operations.

"That sign is typical of their whole approach to safety," Napoleon says. "They try to convince you they're doing a good job so you won't force them to take any real action. You just about have to have someone killed before they will change that number."

Napoleon and his fellow workers at the preparation plant face many of the same health and safety hazards that are found underground. The giant coal crushers, shakers, washers, blenders, and dryers make so much noise that many workers become partially or completely deaf before they reach retirement age. Many also suffer from black lung disease because of airborne coal dust.

Although the roof of the six-story plant is not likely to cave in, workers may be seriously injured by falling objects such as loose coal, tools, or construction materials. Two years ago, Claude Chapman's brother, Michael, directed by his foreman to do clean-up work in a coal dryer, was disabled when coal which had built up on the dryer's walls fell on his neck and back. Even after the company agreed that no one would work inside that dryer except in extreme emergencies, another brother, Billy, was told by his boss to do routine work in the same area. He had to risk being fired for insubordination in order to protect himself.

As in a coal mine, unnecessary hazards are created in the plant when maintenance is put off until it can be done without affecting production. Leaks in water lines may be stopped up, instead of being repaired, and spilled coal may be allowed to pile up, resulting in injuries from slipping and falling.

In the winter of 1977, this approach to housekeeping and maintenance created a safety crisis in the Gary plant. Management had failed to anticipate that an old boiler in the plant might burn out, and when it did, in November, 1976, they were not prepared to fix it. In December's cold weather, pipes that carried water through the plant for heat cracked and burst. Water poured out, mixing with spilled coal and forming icy puddles. The workers were left with no heat through one of the worst winters in West Virginia history.

Napoleon and Claude became so concerned about conditions in the

plant that they gave the superintendent an ultimatum: if the company didn't take strong action within two weeks to start a clean-up program, the local union would shut the plant down. In response, the company agreed to hire ten new workers, who would work solely on cleaning up puddles and coal piles.

''Whenever there's a breakdown, they call it bad luck,'' Napoleon says. ''But usually it's bad management, where they don't plan ahead. They always hold it over us that we need management to run the plant, but this kind of thing makes you wonder.''

After the clean-up program began, officials at U.S. Steel's headquarters in Pittsburgh were asked to allow reporters for the UMW Journal to take a tour of the plant, but they refused. An official explained that ''we have had a lot of problems down there because of the winter, a lot of pipes busted, things like that, and this just wouldn't be a good time for you to visit.''

The superintendent in Gary and his labor relations director weren't willing to help arrange a tour either. In a long conversation at their offices, they talked instead about how there weren't many safety or contract problems at the plant, thanks to their own progressive attitudes.

"We teach our foremen to make sure our workers have the best protection we can give them, and that they get every single benefit they're entitled to under the [UMWA] contract," the labor relations man said. "It's much better to take that approach than to try to lean on them or put something over on them, because you always pay for that later in some way."

Told of the company's claims, Claude and Napoleon grinned.

"At first you think, this guy dresses in nice clothes, he smiles a lot, got a good personality, he's just a human being like you and me," Claude said. "So you expect the guy to be reasonable. But then you go in that office there to deal with him as a representative of U.S. Steel, and it's a whole new ball game, I don't care who he is.

"We used to have a foreman down at the plant who was supposed to be a Christian man. He was supposed to go to church all the time and not lie or cheat or anything.

"But I don't know how many times I sat across the table and watched that man lie or do anything possible for U.S. Steel, just to beat

the union out of something in our contract. Buddy, he just laid the Bible down when it came time to pick up that contract.''

Napoleon's elderly uncle, Willis, was telling visitors about his youth when a knock on the front door interrupted his story. ''Come in,'' he said, and in through the screen door bounded an Oriental teenager with a large carton under his arm.

''Are you the man of the house?'' the boy asked, and Willis assured him he was. ''Well, I'm a lay missionary from the Unification Church, and I'd like to know if you'd like to buy some peanuts to support our work.''

Willis had never heard of the Unification Church or its Korean leader, Reverend Moon. Moon's ''missionaries'' are a familiar sight in the airports and busy streets of major cities, but not in the remote communities of Appalachia.

''Well, I might be able to buy some of your peanuts, but first I've got to know what doctrine do you preach,'' Willis said.

''Sir?''

''What doctrine do you preach?''

''I don't know what you mean, sir.''

''Well, how do you baptise?''

''I've never been baptised.''

''You've never been baptised? And you're supposed to be a missionary?''

It was time for Willis to do a little missionary work of his own.

''Now, I don't know about your church, but we are what some folks call 'Jesus Only' people. We don't baptise in the name of the Father, Son, and Holy Ghost, because we believe they are all one and the same person: Jesus. So we baptise in the name of Jesus. Now if you don't baptise that way, you preach and teach false doctrine, and you are going to go to Hell.''

The boy stammered out a quick reply.

''Sir, maybe sometimes instead of having so many divisions among people like you're talking about, Jesus just wants us to love each other. Maybe Jesus is more concerned with that than with what you call 'doctrines.'''

Willis had heard that line before.

''Jesus says you must be born again, and I don't care how good you are, I don't care if you don't drink or cuss or mistreat your neighbor, if you ain't born again, that is of the water and of the spirit, you are going to burn in Hell.

''And there is only one way to be baptized, and that is the way the apostles were, in the name of Jesus. It's just like this: there's only one way to get from here to Chicago, let's say it's Route 52. You get on that road and you'll get to Chicago, and if you get on any other road you won't. That's the same way it is with the road to Heaven.''

Realizing that Willis wasn't going to buy any peanuts, the young salesman abruptly said goodbye and bolted for freedom.

''That was a strange young man,'' Willis said as the door slammed. ''First time we've had one like that around here. Now Jehovah's Witnesses, they've come around here before. They've got some strange doctrines, they sure do. I tear 'em up, I just tear 'em to pieces.

''Not boasting now, God forbid, but I see that I'm so far ahead of so many people. I go to Bible school every Thursday night, and, well,

WILLIS MARTIN

knowing that I was baptised in the name of Jesus, and not in the name of the Father, the Son, and the Holy Ghost, it just comes to me like A-B-C-D.''

He paused for breath.

''You know, there ain't many people left that holds up for the right way. We're the fewest people on earth. People have strayed and strayed and strayed. You take this last week, how hot it was and how long it's been hot, and the drought you've been hearing about on the radio, and all the earthquakes—look what a hard winter we had.

''These things aren't going to get better, they're going to get worse, until Jesus comes again and sets up a new order. We're looking for Jesus to come and get his people, the people that is ready for him.''

As he spoke, Willis' face filled with joy, like the face of a small child who giggles uncontrollably when tickled.

''I'm so glad that I've been taught the true way and not some false way,'' he said. ''Because if I would go to Hell, I'd be tormented and there'd be no end to it. But if I go to Heaven, I'll be rejoicing, I'll be *rejoicing,* and there'll be no end to it.

''I'm so glad,'' he said. ''Yes, I am, I'm so glad. So glad. So glad.''

Napoleon remembers that his late father was as religious as Uncle Willis. ''On Sunday morning, he made me go to church, there wasn't any choice about it,'' he says. ''But I don't go too much anymore. I consider the union is the work I do for the Lord, making things better for people where they're living now. You take Uncle

NAPOLEON MARTIN

Willis, the union got him a pension and compensation benefits for his black lung, took care of him in old age. But now he doesn't even keep up his union dues, which is only a dollar and a quarter a month since he's retired.

"Ever since I was little, the people in power, like these coal companies, they wanted you to think, 'We don't have to change anything. We don't have to fight for ourselves. Don't worry, the Lord will take care of us. The Lord will take care of us.'

"Other people might feel different," he says, "but I feel like maybe I'll just give Him a little help taking care of us."

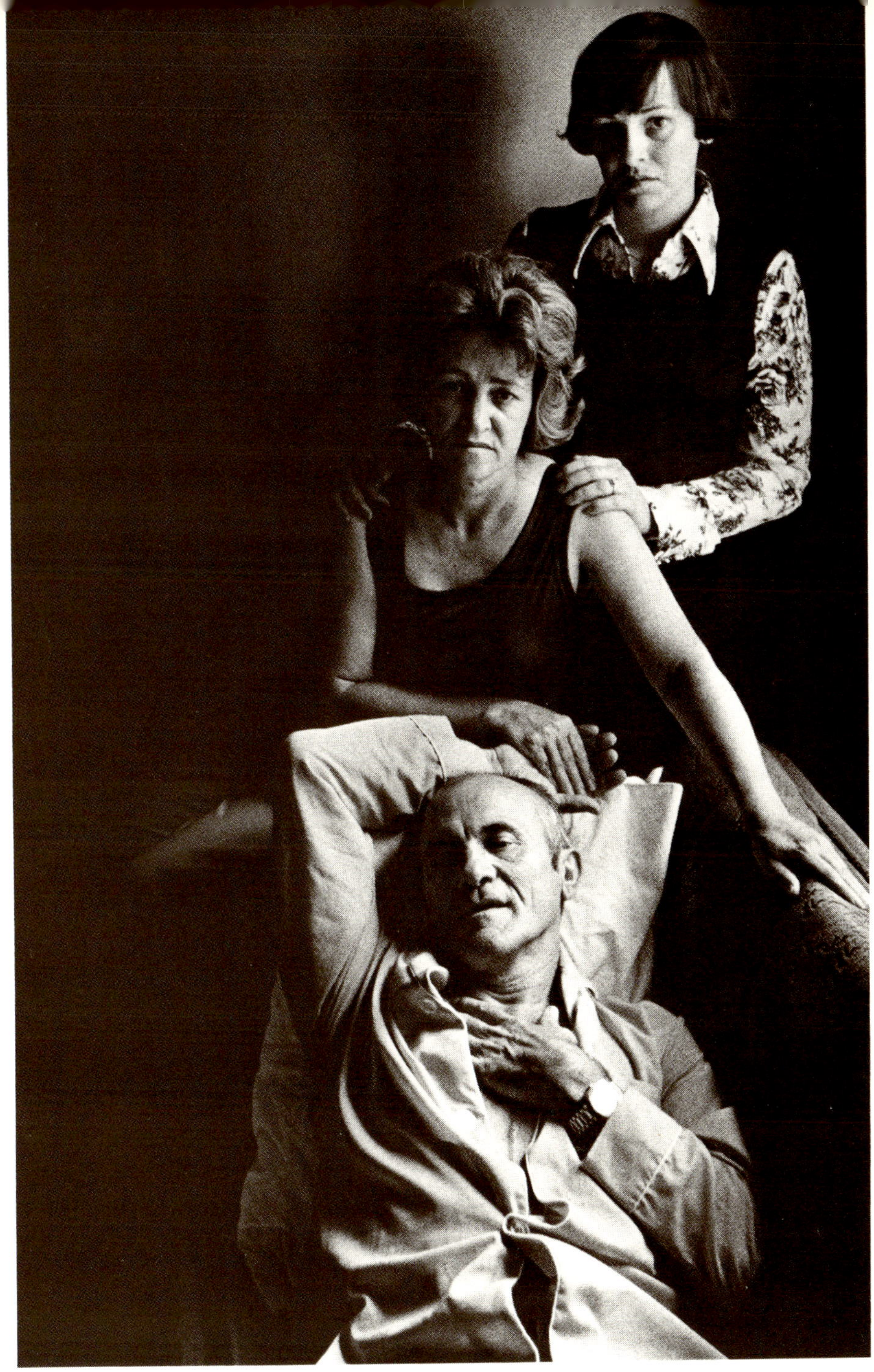

Just Like A War

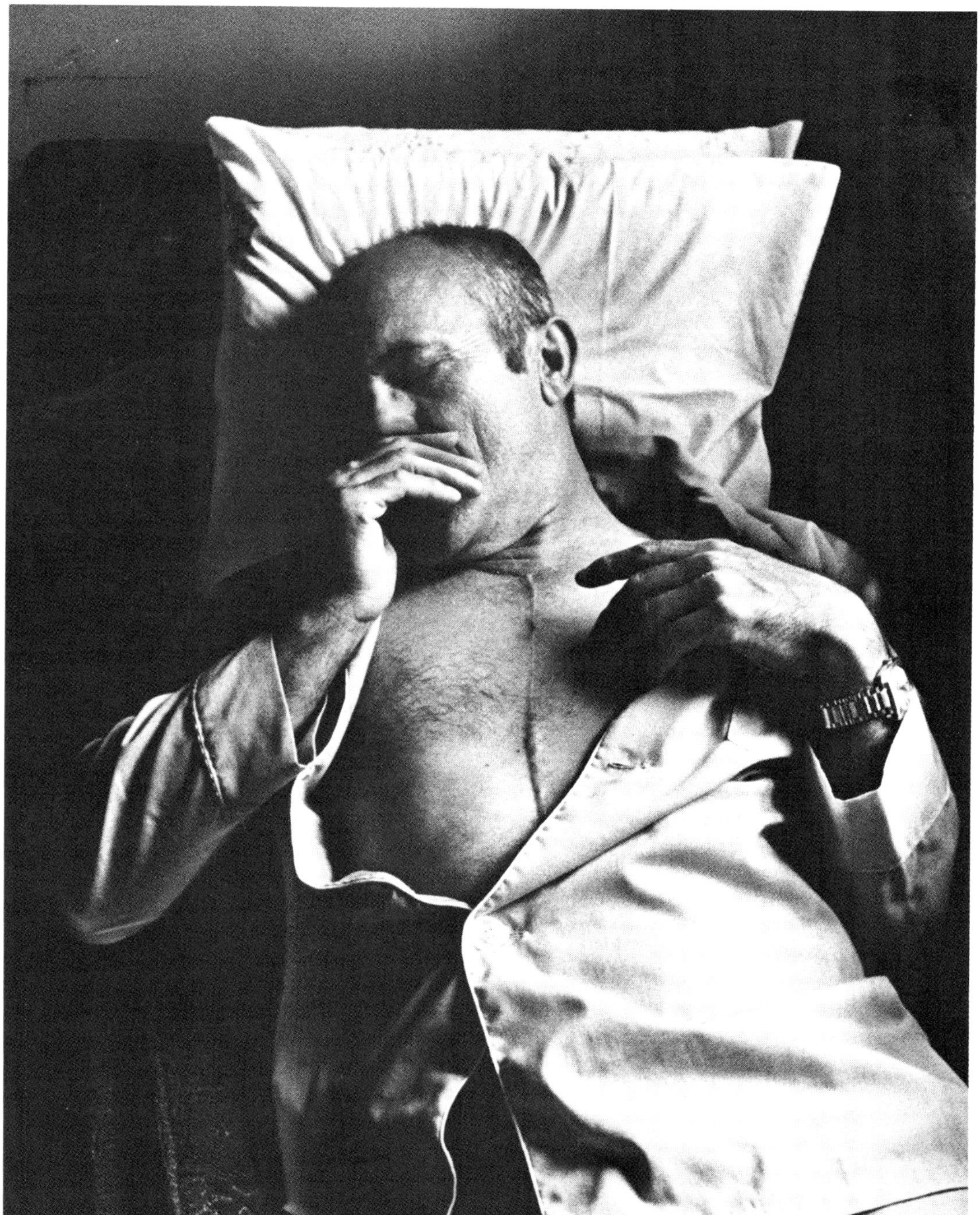

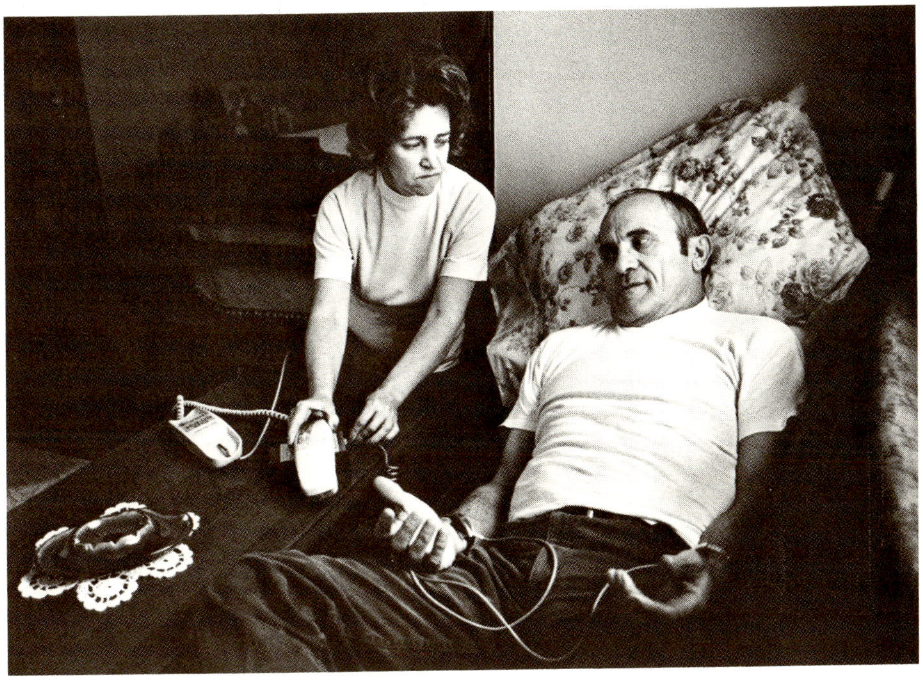

"If I had known then what I know now, I would have never gone near a coal mine," Elmer Lockhart says as he lies back on his living room couch. Wearing a bathrobe although it is the middle of the day, he props up his head with pillows to make his breathing easier.

It has been an especially rough day, he explains, because he couldn't sleep at all during the night. He tried to doze after watching a ball game on TV, but his lungs were too congested. He sat, thinking, reading, waiting, until the inevitable coughing spell came. For several minutes each hour, the pressure in his throat and chest took control of his body, as though trying, not just to clear his lungs, but to cast out the whole devilish disease which plagues him. Then the congestion slowly built up again.

In the morning, he tried to take out the garbage, but the cold winter air made him start coughing. Dust in the summer causes the same reaction.

"I talked to my neighbor for a minute," Elmer says, "but then I had to come right back in. It's kind of hard for him to understand my problem, you know, since they don't have too many coal mines up here in Cleveland."

Elmer and his wife, Gladys, don't have to be asked twice to tell their story—how he got black lung in Amherst Coal Company's Paragon mine in Logan County, West Virginia, how the family had to leave their home when he was fired and blacklisted from the mines for filing for state workers' compensation, how he had to work until he collapsed because he couldn't get federal black lung benefits.

As they talk, a special heart monitoring device sits on the coffee table. Gladys can hook it up to the telephone if Elmer has severe chest pains, and within sixty seconds, a local hospital will tell her whether he is having a heart attack.

"Ever since his black lung got bad, I've had to be the man and the woman both," she says. "Black lung—it's hard to express how much of your life it takes over."

When they talk about the past, it is as though it all happened just yesterday. As they describe how their daughter, Pat, had to give up her college scholarship twelve years ago to help the family move to Cleveland, their voices and faces are filled with a bitterness that shows the wounds are still open. They have been hurt by the loss of Elmer's health, by the separation from their relatives and friends, by the lack of income. But even more, they still can't quite believe that the company, union, and country Elmer served so well could have let them down so totally.

Work hard, and you'll be rewarded. Your life is what you make it. You never get something for nothing. Do right by others, and they'll do right by you.

Growing up in southern West Virginia in the 1920s and '30s, these were the principles that Elmer and Gladys Lockhart learned from their parents, teachers, and ministers.

Elmer was raised in a family of nine children on a farm where, "by the time Mother put you out of the cradle, there was a job waiting for you." The Lockharts grew their own food, raised their own animals, cut their own lumber, and still found time to help their neighbors.

"We'd go to church on Sunday and somebody would say, 'There's a sister or there's a brother whose fields are all in weeds, and we need some help for a few days,'" Elmer recalls. "Next night, my dad would say, 'Come on, boys, we got to go help.'"

To pay for necessities they didn't produce themselves, such as clothes, tools, and an automobile, Elmer's father also put in a full shift each day in the coal mines. Gradually, the Lockharts, like other southern Appalachian families, became more and more dependent on their mining income.

"The younger people, like my sisters, started moving into the coal camps, living in the company houses and buying their food from the company store," Elmer says. "It got to be where it wasn't worth it for my father and mother to take care of the farm."

Elmer quit school after the eighth grade and then went to work in the mines himself, loading coal by hand. It wasn't long before he become involved in the struggle to build the coal miners' union.

"A working man just about had to be for the United Mine Workers," he says. "I'd say it was slavery in those days—the company made you work even if you were sick. They didn't care if it was bad roof or dust or water; you had to work or you were fired.

"We had the union at our mines, so then we tried to help these other places where they couldn't get the union. Buddy, in those days it was just like a war. The deputy sheriffs was paid by the company. The state police was helping the company. We had regular patrols trying to cut off their food supply or cut off the trucks that brought their weapons. I saw several men lose their life fighting for the union. But we had to do it."

During World War II, Elmer served in the Navy in the Pacific. It was his first chance to see how people outside the coalfields felt about miners.

"I was out on this destroyer when we got news that John L. Lewis had the miners on strike for a raise," he remembers. "Right away I was for them, because I knew everybody else was getting rich off the war. The companies were making millions, and the guys in the steel mills and the shipyards were doubling what the miner was making.

"But then these sailors from Brooklyn, New York, or California

someplace started saying, 'Oh, these stupid miners, always striking for something, and here we are at war.' Like miners was low-down or something. Well, I couldn't take that. I'd just reach up there and grab that man with my coal miner's hand and pull him toward me and I'd say, 'Hey doc, you ever work in one of them?' I'd say, 'Close those lips, baby, 'cause you're talking to one of those dumb coal miners right here, and I'll tell you, every nickel those guys are striking for, they deserve it.'''

When the coal companies refused to meet the demands of the striking miners, the federal government took over management of the mines so that coal production for the war effort could resume. The other sailors soon realized that Elmer was right about the miners' demands.

''The government run coal real good and they made money, plus they paid the miners their raise that John L. was asking for,'' Elmer says. ''The coal operators said they couldn't make it if they paid that raise, but the government made it just fine.''

When Elmer returned from the war, he found it hard to readjust to life in Logan County after seeing alternatives to coal mining and coal camp life. He traveled to several northern cities, and finally found work at a General Motors plant in Dayton. But he kept coming home for visits to Logan, to see his family and to court Gladys Hurley.

Gladys was the daughter of a coal miner who died when he was twenty-three years old. Her mother supported the four children by doing laundry for other families, using a washboard and stove iron.

''My mother didn't have to teach us what work meant,'' she says. ''That's all we ever knew. I worked at any job I could find, just to buy my own clothes. We fixed up our house the best we could, but I can remember being able to see sunlight through cracks in the walls. My mother had to raise us the hard way.''

After they were married in 1946, Elmer and Gladys thought about settling in Dayton to raise a family. But before they had made up their minds, Gladys became pregnant with their first daughter, Pat, and Elmer decided to go back to the mines.

''Logan County was more like home to us, a better place to bring up our kids, and we just decided to settle down there,'' Elmer says. ''I worked for GM some again a few years later during layoffs, but I guess that was the time right there when I decided to be a coal miner for good.

"We wanted to have a good home and give our children some of the things we never had, and we figured that was the way to do it. And I guess if it hadn't been for the dust and the way the company treated me, it would have worked out pretty well."

There's one memory that stands out in my mind to tell you about my childhood," says Pat Lockhart. "I represented my school in a parade over in Williamson, and then there was a football game, and they would choose a queen from the girls from all the different schools in the state.

"I was supposed to have three outfits, one to be very formal, one a gown for dancing, and a suit. Now, there was no way my parents could afford to go out and buy all that, with the three kids already. So my mother said she would make the things from scratch.

"We had to wait until payday to have the money for the material, and that didn't come until the day before the parade. My mother stayed up all night sewing. She didn't go to sleep at all.

"My dad came home in the middle of the night, because he'd been working second shift, and then we had to leave at about five thirty in the morning.

"So that day they had the parade and we were riding in the backs of these convertibles, all the girls. And all the parents were taking pictures for their daughters to be able to keep. But we didn't have a camera because we really couldn't afford it.

"Well, I looked down from the car and there was my dad, walking along as fast as the car was moving, snapping my picture. He'd gone into the drugstore and bought a camera.

"That's the way my parents were," she says. "Some people don't take an interest in their kids, and some will give their money but they won't give their time. Mom and Dad gave everything. They would do anything for their children."

Both Elmer and Gladys worked hard to fix up the houses where they raised their family. Elmer would come home from work and put in another shift installing a hot water system or a bath. Gladys was one of those housekeepers who believes Cleanliness is next to Godliness.

A COAL CAMP WHERE LOCKHARTS LIVED, DEHUE, WEST VIRGINIA

"I always took pride in our home, no matter what it was like to start out with," she says. "Even if you don't have something fancy, you can make something out of it. You can make it feel like a home."

On weekends, if they weren't going on an outing with their kids, they would often go for a drive up Rum Creek, stopping on the way to see Harry and Geraldine Chambers. Harry's family had been among the first to settle in the area, and the Chamberses still lived in the original old farmhouse, which stood in front of their fruit orchards and vegetable garden. A family named Jones had stayed at the house for a few nights many years back. They were looking around the area for land, and were grateful for a floor to sleep on. Later, with the land they bought, the Joneses founded Amherst Coal Company.

Next to the Chamberses lived Huey Moore, a local union leader and one of Elmer's favorite drinking partners. On a Friday or Saturday night, they would pick up some "moonshine" or stop at a local tavern to "drink some beers and tell a few big windies." Occasionally such events would take place during the day, under the guise of "fishing trips."

"See, most of the time going fishing was just an excuse to get together and relax on your day off from the mines," says Huey, who still lives on Rum Creek. "I can remember once when I went with a bunch of guys and their kids down to the river to fish. Everybody had their lines in the water but they weren't paying any attention to them—too busy talking and having a good time.

"All of a sudden, one of the kids starts yelling, 'Hey, Dad! Dad! You've got something on your line! Come quick!' And his dad says, 'Shhh. Keep right quiet, son, and maybe it'll go away.'

"That's the way we fished in those days," he says. "Keep quiet, and maybe it'll go away."

ELMER LOCKHART'S FRIENDS HUEY MOORE, TOMMY PROPS

Just as Elmer's father once had worked two jobs to make ends meet for his family, Elmer often worked double shifts for his, spending up to sixteen hours at a time in the mine. He did all the dustiest jobs, running everything from the machines which cut the coal to the buggies and belts which brought it outside.

When he reached his late thirties, he began to notice that he would get short of breath when doing routine jobs like carrying timbers or sacks of the rock powder used to keep down coal dust. At night, he would cough up dust and mucous for hours.

He began going frequently to the hospital at Man, West Virginia, for medication when he couldn't sleep. The doctors there told him that cigarettes might be causing his problem, even though he had quit smoking. They said he had a cold, or the flu, or was suffering from exhaustion.

Amherst and other coal operators denied that coal dust was harmful; one company on Rum Creek taught its workers that the coal dust wouldn't hurt them because the particles were round and would roll right out of their lungs. Although black lung disease, or "coal worker's pneumoconiosis," had been officially recognized in British mines for several decades, American companies did not want to spend the money to keep down the dust with fresh air and water sprays. Coalfield doctors, politicians, and union officials were either ignorant or afraid to speak out.

Elmer suspected that he wasn't getting straight answers from his doctors, and in 1963, he made an appointment with a lung specialist outside the immediate area, in Huntington, West Virginia. The doctor gave him X-rays and other tests, and told him he had an incurable lung disease. ''Silicosis,'' the doctor called it, meaning that he believed it was caused by inhalation of silica dust from rock in the mine.

The Lockharts faced a tough decision. For his health, Elmer knew he should quit the mines. But for someone with his skills, work experience, and education, there were few other jobs in Logan County that would pay enough to support the family, especially as the children approached college age.

He decided to file for state workers' compensation. Although the law did not recognize lung disease caused by coal dust, it did authorize benefits for silicosis. Elmer thought that with compensation benefits, he eventually might be able to save enough money to allow him to accept lower-paying work outside the mines. The state had him examined, and held a hearing about the dust conditions on his job. Then the Lockharts had to wait.

''I was terrified,'' Pat recalls. ''Just before that, I had been sent from the office at school to get one of my girlfriends out of class because she had a phone call. As soon as I told her there was a message, she burst out crying. Her father had been in the hospital, dying from black lung.

''Well, in those days they would display the body in the home, and I went to see him. And I saw my friend's father lying there, he was a young man, and I saw his young children standing around. And then it hit me: 'Oh, my God. Is this what is going to happen to my father? Is this what is going to happen to my family?' ''

Finally, in April, 1965, the letter came. Gladys opened it while Elmer was at work. It said he had silicosis, advanced stage, but that he was considered only fifteen percent disabled and would receive only about two thousand dollars. It was the worst possible news: confirmation of the seriousness of his disease, but not enough money to free him from further exposure to dust.

The next day, Pat stayed home from school to help her mother break the news to Elmer. After Gladys told him about the letter, Pat moved across the room and hugged him.

''I told him, 'Everything's going to be okay,' '' she recalls. ''And Mom told him, 'Don't let this worry you. Everything is going to work out.' ''

A few days after Elmer's compensation award became final, he was fired from his job. The Amherst superintendent called him into the mine office, and gave him a slip of paper. It said his work was ''unsatisfactory.''

Knowing that he had an excellent work record, he demanded that company officials retract their lie. Two weeks later, they promised to change the wording of the discharge. Amherst now said it wanted to ''protect'' him from more exposure to dust.

Elmer knew the company was making an example of him. Few coal miners at that time ever filed for state compensation benefits for silicosis. Many sick miners who would have been eligible did not know what disease they had. Others guessed, but were afraid to file for fear of losing their jobs. Amherst and the other companies which had to finance the compensation fund wanted to keep it that way.

Well-known and well-liked, Elmer was vice-president of his local union at Amherst's Paragon mine. If he could get away with applying for silicosis benefits, others also might be encouraged to try.

His local was ready to strike to get his job back, but he talked his buddies out of it.

"We'll take this through procedures," he told them. "This is what we have a union for."

But to his surprise, the union's district representative in Logan urged him not to challenge the firing.

"The man told me, 'Do you know what you're getting into?'" Elmer recalls. "I said, 'What do you mean? I'm forty-two years old and I can't quit working.' And all he would say is, 'I'm just warning you, you better watch yourself.'"

Despite the warnings, Elmer's local decided to pursue the case. Evidence would be presented to a joint board of two union district representatives and two operator representatives. If neither side gave in, the case would be decided by an arbitrator.

At the grievance hearing, Elmer argued that since the company admitted his work was outstanding, only he should decide when he could afford to get out of the mines.

An Amherst representative, in questioning Elmer, concentrated on proving how unhealthy the mine was.

Wasn't it true that there was dust at the boom where he worked? Wasn't it true that rock dust was heavy in the air along the haulage ways? Wasn't it impossible to see other workers at times because of the dust? If so, he concluded, then why should Elmer be allowed to keep working there in his condition?

No one pointed out that the company had proven its own failure to maintain a healthy workplace. No one demanded that the company protect Elmer by controlling the dust, rather than by firing him. Instead, the two union representatives—appointed payrollers who owed their jobs only to the union president in Washington—agreed to uphold the firing.

Other companies were sure to learn of the case through the Logan coal operators' association, or by checking Elmer's work record if he ever applied for a job.

He could never work in the coal mines again.

Until Elmer was fired and blacklisted, the dreams Pat inherited from her parents were well on the way to becoming true. The day after Amherst took away Elmer's job, Pat graduated from Man High School. She had won a one-year scholarship to Marshall University, and had arranged work for the summer to help pay her first year's expenses there. She was on the verge of achieving her lifelong goal of becoming a teacher.

"I had always been taken by the look on a child's face when they first grasp something they didn't know before," she remembers. "I grew up with the idea that education is the Rock of Gibraltar you hang onto. This is what makes your life richer. And I guess I wanted to be a teacher so I could share it with others."

Pat had made the most of her time in school—president of the Future Teachers of America club, member of the National Honor Society for "A" students, Prom Queen, "Miss Senior," and secretary of her church's Baptist Youth Fellowship. And like the teacher she wanted to become, she helped others make the most of their skills as well.

ELMER LOCKHART WITH SISTERS BEFORE HE LEFT FOR CLEVELAND

She spent hours helping her cousin, Darrell Mutters, with his schoolwork. He was more inclined toward mechanical skills than academic subjects, but she wanted to help him finish school so he wouldn't be "stuck in the coal mines if he didn't want it."

Carol Evans, her closest girlfriend, remembers that "Patty always made me come out of myself.

"It's a little thing, but I can remember one time I went with Patty's family on a picnic, and there was a place to wade there," she says. "I was afraid of the water, and she didn't know how to swim either. But next thing, she was out in the water, and then somehow she got me out there too.

"The water was there to enjoy and she just never let opportunity slip by. She was the same way with people she met, or classes she took, or any kind of activity. She wouldn't let you say, 'No, I won't do it today, I'll do it next time.' She really made people feel important, like they could be more than they thought they could be."

After all the hard work Pat had put into her education, Elmer and Gladys were determined not to let the loss of Elmer's job change her plans to go to college. But without his income, how would they pay college expenses for Pat after the first year, and for the two younger children when they finished high school? How would they make the payments on their house and car? How, for that matter, would they eat?

They decided that Elmer would have to go north to look for a factory job. Gladys and the younger children would stay behind until he found work and a place for them to live.

"My parents sat me down and said, 'Now, you just go on to school, use your scholarship this year, and maybe something will work out to pay for the rest of the time.'" Pat recalls. "But I couldn't do that. Who would take care of my father if he didn't feel well? Who would help him look for a house, and be there to cook for him?"

Pat and Elmer drove to Cleveland, where they stayed briefly with relatives until they found a temporary apartment. They cooked on a hot plate and slept on air mattresses on the floor. Looking for a house for the whole family, they met suspicious looks from landlords who wondered "what the relationship between these two hillbillies really was." To get a job at a Westinghouse factory, Elmer had to lie about his health.

They felt humiliated and lonely.

"It was so painful to see my dad going through this," Pat remembers, "having to leave what he had worked so hard for and start from the bottom all over again. But it was either that or starve."

They finally found a small house to rent, with three rooms that "really wouldn't have made one big room." They quickly sold their house in West Virginia, without getting anything for the years of labor they had put into it. Gladys and the younger children came to Cleveland.

While Pat worked as a secretary for the American Greeting Card Company, Elmer worked full shifts at the factory and then extra hours at other jobs. The coal dust in his lungs was gradually making his breathing more difficult, which in turn put added pressure on his heart. He would work until he collapsed, spend a few days at home or in a hospital or clinic, and then drag himself back to work. There seemed to be no way out.

Despite Pat's tutoring efforts when they were in school, her cousin, Darrell, did go to work in Amherst's mines. He started on the "hoot-owl" shift (midnight to eight a.m.), and unlike most new miners, he deliberately stayed on that schedule even when he had the seniority to bid on jobs on other shifts. Since those hours were primarily devoted to maintenance and repairs of broken-down equipment, instead of coal production, he was exposed to less dust. He had seen too many older miners, including his own father, looking "pale and shaky like they were walking around dead" because of lung disease. It was inconvenient to sleep during the day or evening when his family and friends were awake, but he figured he could stand a little inconvenience if it meant protecting his health.

Darrell's fears of the dust grew when he heard other miners talking about meetings of the new Black Lung Association. In 1968, disabled miners and widows, with the help of young organizers who had first come to Appalachia during the War on Poverty, established the BLA to lobby for black lung compensation and dust control programs.

At first, their meetings were attended by only a few committed activists, but then more people began to come as word spread about the "black lung doctors." These three doctors came to the meetings with specimens of miners' lungs that had been clogged with dust. They would hold up samples of blackened tissue and warn their audiences: "This is happening to you, or to your husband. Whatever you have been told in the past, the truth is that coal dust will kill you."

In January, 1969, the BLA convinced members of the West Virginia legislature to introduce a bill to provide state workers' compensation benefits for black lung itself, as well as for silicosis. The group hoped to win benefits for miners no longer able to work and for their widows, and to provide the companies with a financial incentive to control the dust.

The proposed legislation immediately ran into strong opposition from the coal operators, who traditionally financed the election campaigns of many of the lawmakers. The miners felt they had only one way to counteract the companies' money: they stopped work. For three weeks, forty thousand West Virginia coal miners shut down coal production. The union's international and district leaders were furious—the strike was illegal and unauthorized, the work of "outsiders" and "radicals." But experienced local union activists like Huey Moore and

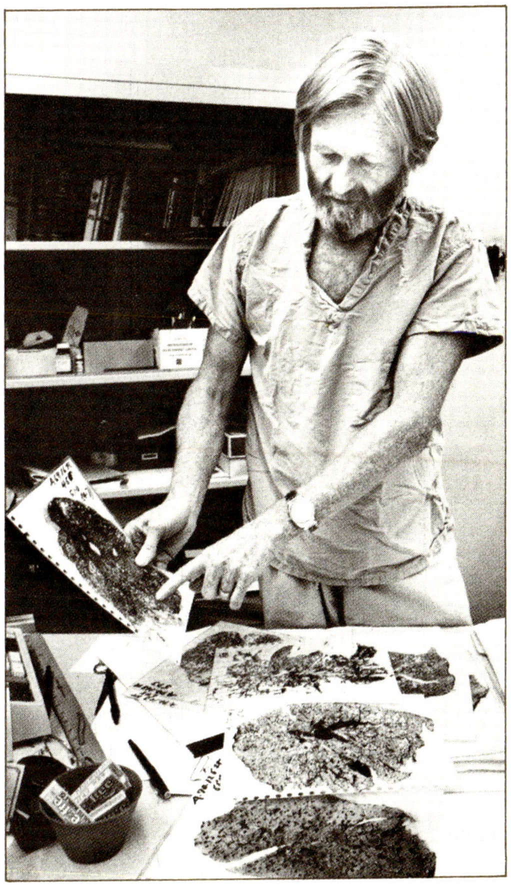

BLACK LUNG DOCTOR DONALD RASMUSSEN

BLACK LUNG VICTIM, SOUTHERN WEST VIRGINIA

newcomers like Darrell Mutters and his brother, David, worked together to keep the pressure on the legislature.

Although they were losing valuable wages, the strike was a joyous experience for many of the participants. They were landing solid punches, not only against the lung disease and the operators, but also against despair, against the feeling that coal miners just have to swallow what is given to them.

''I did run into some of the older men who had sort of given up, been beaten down so long by these companies that they didn't think we could do no good,'' recalls David Mutters. ''But I told them, 'Listen, man, I might have to work in this thing another twenty years, and I don't want to work in these conditions.'

''I was tired, staying on the roads day and night, helping out on the picketing, but it was worth it. We had some momentum, buddy. We were pretty strong during that time.''

With hundreds of miners watching from the galleries, the legislature passed the state black lung bill in March, 1969. Then the BLA shifted its attention to the U.S. Congress, which was considering a federal coal mine safety law in the wake of the Farmington, West Virginia, explosion that killed seventy-eight miners. With a nationwide wildcat coal strike an obvious possibility if no action were taken, the

1969 Federal Coal Mine Health and Safety Act was passed, providing monthly benefits for black lung victims or their widows, as well as specific standards for dust control.

In February, 1970, less than two months after the new law was enacted, Elmer began the process of applying for benefits. If he were approved, he would be able to quit his factory job. His disease would never be cured, but by putting less strain on his heart and lungs, he could probably ease his discomfort and prolong his life.

No sooner had the government's promise to Elmer Lockhart been made than it was broken. The job of carrying out Congress' mandate fell to the Nixon administration, and that was where the black lung program's downfall began. Nixon had opposed the 1969 health and safety law, claiming that it would put too great a burden on the coal companies. Although black lung benefits would come from the federal treasury, eventually the operators would probably have to assume some responsibility, and that could cost corporate stockholders millions of dollars.

Once the law passed, Nixon's opposition to it took a new form. Elmer and other miners found that the regulations the administration reluctantly established for black lung benefits were so restrictive that "you almost had to be dead to qualify."

The administration was aided in dragging its feet by the disagreement among doctors on a precise definition of black lung. There might be other diseases contributing to a miner's breathing difficulties, or he might have related health problems like heart disease, and no test could measure only the effect of coal dust. If a miner was given one kind of test by a doctor who said he had the disease, he could always be sent to another doctor for another kind of test, or the test results could be "reinterpreted" by the government's doctors.

During the first seven years after the law was passed, about two hundred twenty-five thousand miners were found by the government to be totally and permanently disabled by black lung—more than four times the number of Americans who died in the Vietnam War. One hundred forty thousand widows of black lung victims also qualified for

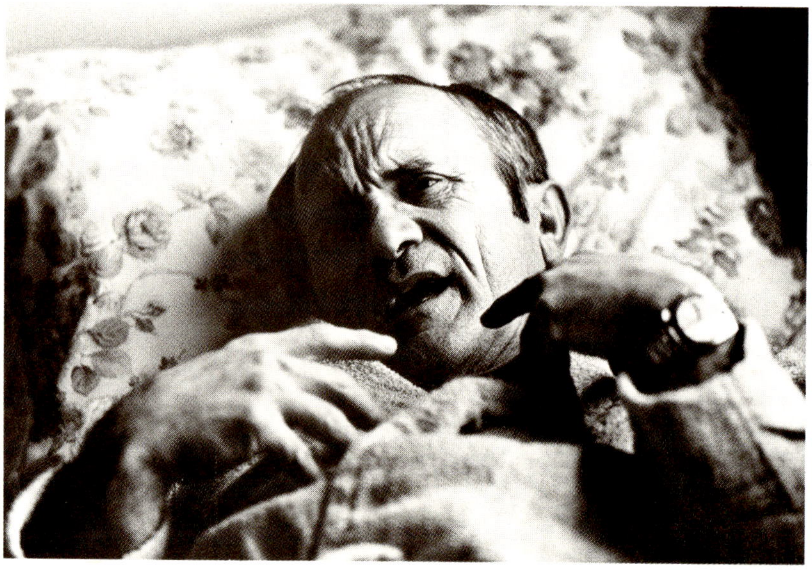

benefits. Yet about three hundred thirty-five thousand people—about half the total applicants—were not awarded benefits. Some of them did not have legitimate claims, but many were simply unable to fight their way through the government's complicated system for determining eligibility.

On March 16, 1971, the first decision on Elmer's claim was issued by the Social Security Administration, the agency then in charge of black lung benefits. The government admitted that he had black lung. But it argued that since he was still working to support his family while he tried to get benefits, he must not be disabled by the disease. His application was denied.

Six days later, Elmer collapsed on the job. He was in the hospital for two weeks. A year later, after several more setbacks, he finally had to quit work.

Elmer was given X-ray after X-ray, and at least four doctors concluded that he had black lung. Using different tests, four other doctors also told him he had it.

Elmer's claim was appealed and reconsidered several times, but still was denied. Pat became part lawyer, part investigator, as she gathered new evidence and prepared for the next round of arguments.

"Getting these benefits that my father had coming to him became part of my daily life," she says. "I used to lie in bed at night, thinking about what calls I could make the next day, or what letters I could write. I thought about going back to school, but I couldn't because I had to be available to meet with someone at the hospital or the Social Security office or make another call."

Month after month, Pat worked on the case, tracking down Elmer's medical records from ten years before, locating the transcript of his discharge hearing, studying federal black lung regulations, writing to members of Congress, taking Elmer for more tests.

"The whole testing process was killing my father as bad as the disease," she says. "I sat outside the door at the clinic one time while they gave him a breathing test. The man was shouting, 'Deeper! Deeper! Blow! Blow! Blow!' My dad's face went red, and then it went completely pale. This is not justice; this is torture. This is not what you put a sick man through. And I will always wonder what all those X-rays have done to him.

"Then I had to take him to the welfare office to apply for food stamps, because all he had coming in was a little money from veterans' disability. To him, he thought of it as charity, and it hurt him."

For several months, Social Security's main office told Pat that they had lost Elmer's file. When she called to demand to know what they were doing to find it, they didn't seem very concerned. It couldn't have been thrown away, they assured her. It must be around somewhere.

In September, 1974, more than four years after Elmer filed his black lung claim, Social Security scheduled a hearing for him with an "administrative law judge," the first bureaucrat who would hear his case in person. The law judge refused to let the Lockharts see their file ahead of time and barred Pat, their best advocate, from the hearing room.

A year later, the judge issued his decision. He said he had disregarded the results of a key breathing test and had given "little weight" to mere "observation and opinion" contained in examination reports by Elmer's doctors. Once again, Elmer was denied benefits. In the mean-

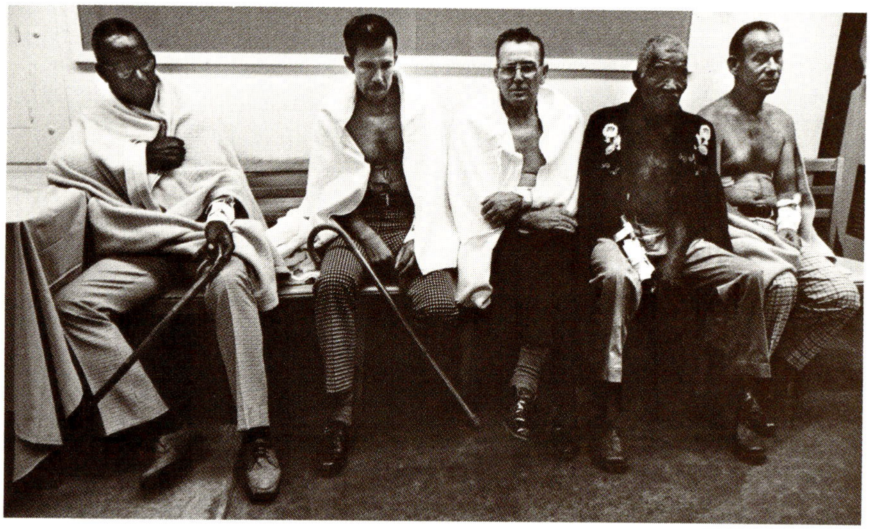

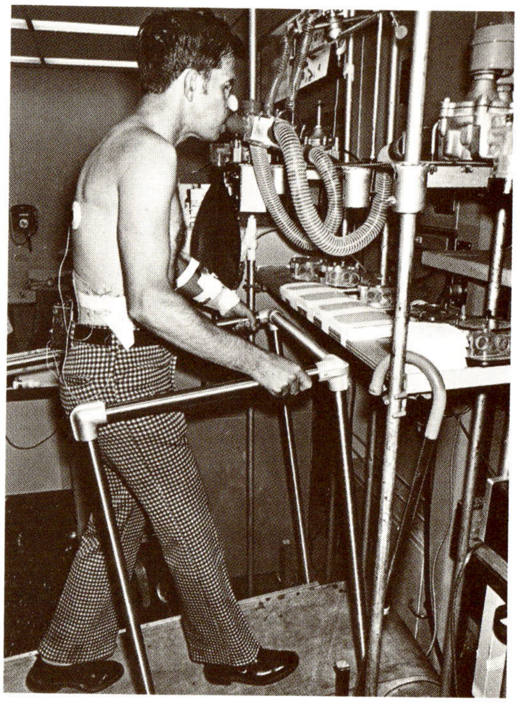

time, he had open heart surgery, and a biopsy confirming his lung disease.

The Lockharts had only one more chance: a hearing with Social Security's "Appeals Council." It was scheduled for October, 1976, thirteen months after the law judge's decision, in Washington, D.C. Pat and Gladys would have to represent Elmer, who was too sick to make the trip. They would have to pay their own way.

While thousands of disabled miners like Elmer were being turned down for black lung benefits, the government was also breaking its promise to control dust in the mines.

In theory, the 1969 federal law was supposed to provide strong penalties for operators who ran dusty mines. Individual miners would wear samplers to measure dust on their jobs, and federal inspectors would make surprise visits to check dust levels. Companies could be fined up to ten thousand dollars for each violation of strict dust control standards.

But here again, the Nixon administration refused to convert the law's theory into practice. The companies, instead of the miners, were put in charge of taking dust samples, and they took full advantage of the arrangement. Officials of Consolidation Coal Company, the nation's second largest coal operator, were found guilty in federal court of tampering with air sampling devices and falsifying sampling records. Most miners believe that other companies have just not been caught.

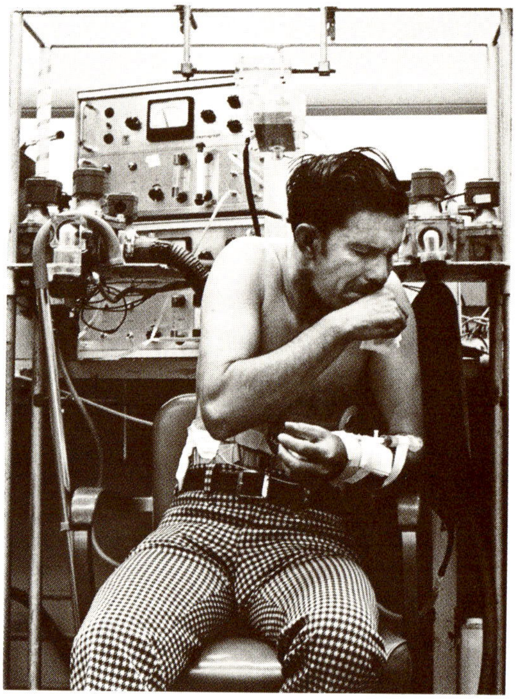

When Amherst was discovered exceeding dust standards, it was fined an average of only forty-six dollars per violation. Inspectors routinely violated the law themselves by notifying the companies before they came, and virtually never used their legal power to shut down dusty mine sections for repairs.

"With all the high-powered machines, the dust is actually worse than it was in the old days of handloading," says Lee Hipshire, a local union health and safety committeemen who used to work with Elmer at the Paragon mine. "Most of the inspectors that are supposed to do something about the dust once worked for Amherst or one of the other companies. They sit up in the mine office, drink pop and eat the company's cheeseburgers; they get along fine. If they come under-

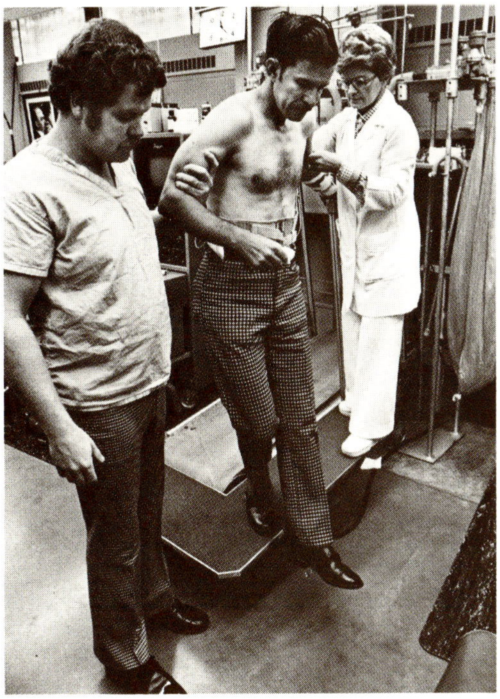

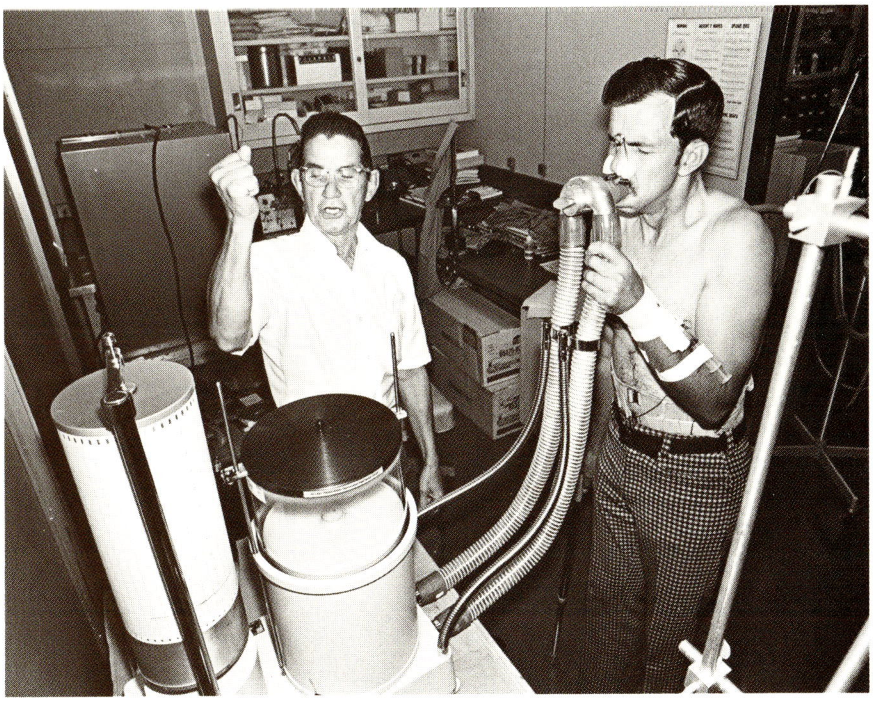

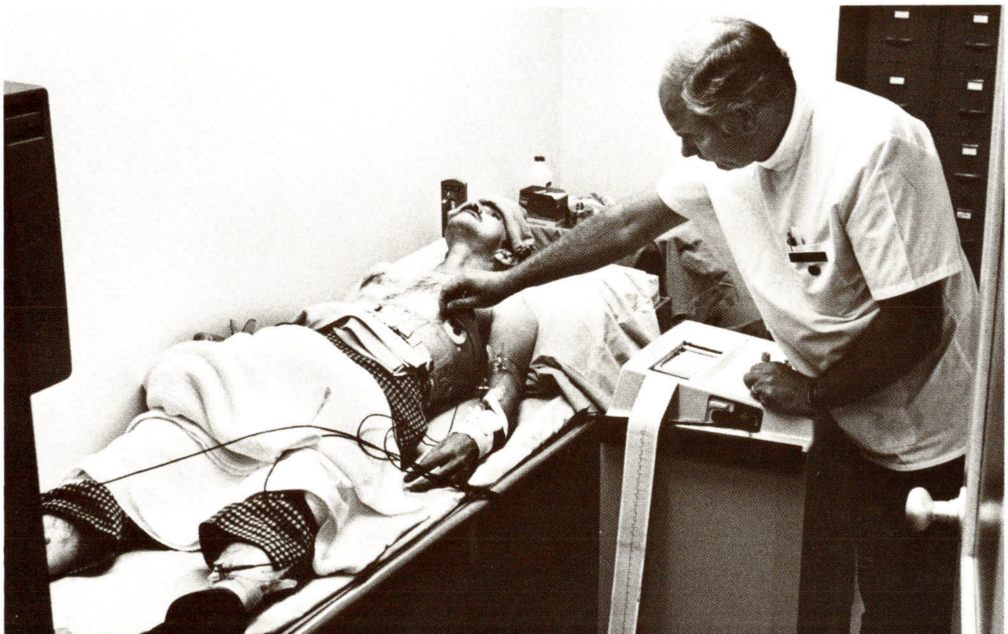

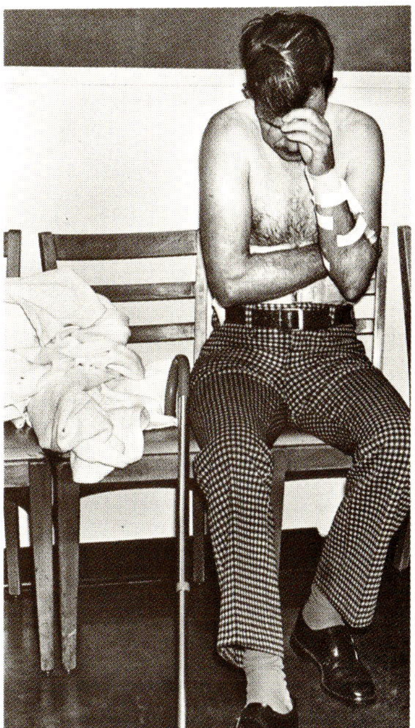

MINERS UNDERGOING TESTS AT CLINIC, BECKLEY, WEST VIRGINIA

ground, they might cite a man for not having his safety glasses on, but walk right by a machine that's been allowed to get clogged up and isn't putting out any water spray.

"Gag on a gnat and swallow a camel is what they do," he says. "We pay their salaries with our taxes, but most of them sure don't work for us."

With the weaknesses in the federal program, the miners at Paragon have to try to enforce dust controls themselves. Under the UMWA contract, each miner has the right to refuse work in excessive dust or other conditions which cause an "imminent danger" to his or her health or safety. But few workers want to go out on a limb by invoking that right. Instead, they call on the local union health and safety committee to present their grievances to the company.

The company, according to the contract, is supposed to give its superintendent and foremen the authority to quickly settle disputes at the mine. Amherst, like other companies, ignores that obligation. To discourage miners from filing grievances, it stalls by claiming that higher-ups in the company must make the final decisions, or by forcing the union to take even the most obviously valid complaints through a long arbitration procedure.

"I've been on our local's safety committee," says Elmer's old friend, Huey Moore, who still works at Paragon, "and I can tell you this: it's not much easier to get the company to do something now than it was when Elmer was here. You can report a grievance until you're blue in the face. You can tell them it's too dusty, it's too dusty, it's too dusty. But all that talk is no more than spitting in the creek.

''Now, I need to work and I like to work,'' he says. ''I've got a family to support. But if it comes down to protecting your life when there's too much dust or an unsafe condition, you just have to shut them down until it's fixed. The companies will do whatever they can get away with unless there's a penalty that makes it too expensive.''

Depending on the issue and the size of the companies' coal stockpiles, some strikes are successful and others are not. But regardless of their outcome, they take a large toll on the miners. Lee Hipshire, who at age forty-six has worked for Amherst for twenty-six years, needs a steady income to support his family of six and to start building up savings for an early retirement from the mine's dusty environment. His wage rate is about sixty-five dollars per day, but his actual earnings are cut substantially by strikes and by frequent illness.

Often, Lee can't sleep at night because of lung congestion. He sits on an old couch, watching the coal burn in his fireplace, wearing his favorite tee-shirt with the state emblem and the slogan, ''Almost Heaven, West Virginia.'' It is a lonely time; the TV stations stop broadcasting and Lee, unlike Elmer, is not interested in reading. The night's silence used to be broken now and then by chickens that were raised in a long building next to his old company house. But the lack of housing and the increase in the number of miners of Logan County have convinced the owner to start converting the old chicken coop into apartments for people.

In order to qualify for UMWA pension, Lee will have to try to work another nine years. He knows there is a good chance he won't make it.

The Appeals Council hearing room where Elmer's case would be decided was, like a courtroom, designed to make those who sat in judgment seem larger-than-life. The three council members would sit on a raised platform in padded, high-backed chairs with a flag behind them, while Pat and Gladys sat below in simple wooden chairs behind a small table. Pat and Gladys were told that when the three men entered the room, it was customary for everyone to stand, and to remain standing until the council members sat down.

When it was time for Pat to present her father's case, she quickly ran through the facts—the doctor's report in 1963, the state silicosis award, the firing, the medical exams since then, the mistakes by the administrative law judge.

She paused. She had never had a chance to tell anyone from Social Security how she felt about the way her family had been treated. Now was the time.

''I wonder if any of the three of you has ever been in a coal mine, or even traveled to the coalfields,'' she said. ''I wonder if you have seen a man die a little bit every day. This is what my mother and I have seen. We have seen politics played with this program, but does anyone ever stop to think that they're playing politics with other human beings?

''I heard the president on TV the other night talking about how we're going to have to mine more coal. And while he was talking, right away I thought of my father, and I said, 'Is this what the president has in mind?' ''

Then it was Gladys' turn.

''I've seen my husband go from a strong, healthy man to a man you

COAL MINERS AT WORK, LOGAN COUNTY, WEST VIRGINIA

have to take care of like a baby,'' she said. ''He's vegetating; he's not living. He's fifty-four years old. He should be in the prime of life.

''We shouldn't have to be here today. But we came because we want my husband to get his benefits before he dies. We don't want to have to produce his body before he gets what he's earned.''

When Pat and Gladys were finished, they waited for questions from the three council members. There weren't any.

The Lockhart's friends, Geraldine and Harry Chambers, weren't lucky enough to have someone like Pat to represent them. Harry, who was Lee Hipshire's father-in-law, worked for forty-two years in the mines, but when he applied for black lung benefits, he was turned down. In 1975, five years after he filed, the first hearing was granted on his claim. By then, black lung had already killed him.

After Harry died, Geraldine put her claim for widow's benefits into the hands of a lawyer in Logan. She tried to attend a hearing on the case, but was sent home by the lawyer, who said he would handle it. She called his office every few months, but his secretary always told her he was busy. ''These things take time,'' the secretary said.

Finally, Geraldine contacted her representative in Congress, whose staff made some calls about the status of her case. In early 1978, she learned from a black lung program official that her claim had been approved. A review of autopsy records showed that Harry had black lung.

The lawyer received twenty-five percent of Geraldine's back benefits (or more than five thousand dollars) directly from the government under the terms of an agreement every client must sign before the lawyer

LEE HIPSHIRE

I EE HIPSHIRE AND SON
AT HARRY CHAMBERS' GRAVE

GERALDINE CHAMBERS

will take a case. She says she is not sure that he did anything to help win her claim; he never did contact her, even when the government's decision was announced.

Not knowing whether he or his wife would ever receive black lung compensation, Harry had agreed before he died to a deal proposed by Amherst. For a nominal fee, he let them have his land and knock down the old Chambers home and orchards in order to build a supply yard for their mines. In return, they gave his wife just enough space for a mobile home where she could live out her days on her Social Security income of two hundred dollars per month.

When Harry died, his family's past hospitality to the Amherst founders, the Joneses, and his years of service to their company were not forgotten. An Amherst employee who worked in the supply yard was sent across the road to the Chambers family graveyard, on company time, to dig Harry's grave.

A week after the Appeals Council hearing, Elmer received a letter from Social Security. "If you disagree with this decision . . ." it began. It continued with instructions on how to appeal. Elmer had gotten denial letters before which began the same way.

"Turned down again," Elmer called to Gladys. "Turned down again."

He continued reading to the second page. *The Appeals Council,* it said, *has decided to award you benefits.*

There were tears in his eyes as he phoned Pat. When he said they had a letter from Social Security, his voice was so choked with emotion that she thought they had been turned down again.

Finally he got it out. "We won! We won!"

"I couldn't believe it," Pat recalls. "At first I was just so happy for my parents. I wanted to stand on a mountaintop in Logan County and scream our victory. I usually keep our problems to myself, but I called everybody I know to tell them we finally won."

For a few days, excitement pushed bitterness into the background. Then the Lockharts began to realize that very little had changed with their victory.

Elmer would receive back benefits, but the seven years in which he could have best spent the money were gone.

Social Security had been convinced to recognize his disease, but that wouldn't affect tens of thousands of other miners who could not pursue their cases for as long as the Lockharts.

And Amherst Coal Company, which had started Elmer's problems, would be able to leave it to the taxpayers to finance his black lung benefits.

"I think about the men who run Amherst, of course I do," Pat says. "The men who decided to fire my father, they went back home the night they did it, and their families didn't have to worry. They didn't have to fetch for themselves. Their children didn't have to give up their education.

"Black lung destroyed my father's life and his family," she says. "But to Amherst, he was just another coal miner."

LEE HIPSHIRE AND SON, HARRY LEE

Lee Hipshire's oldest son, Harry Lee, was a miner for two years. The machine he operated did not come with automatic water sprays to keep down the dust. Lee and the other local union committeemen asked Amherst to install the sprays, but the company refused.

Lee explains that the committee can only go so far without united backing from the miners.

"Some men are afraid that the company will get on them until they find some way to fire them, or put them in a mud hole to work," he says. "They've seen it happen before, and they don't want it to happen to them."

Harry Lee said in the spring of 1977 that he didn't expect dust conditions to improve.

"My throat gets pretty dry eating all that dust," he said. "I don't know. I don't like it, but there's not much you can do about it. I've got training working on diesels, and I could get a job in Huntington or somewhere, but I don't want to leave this area.

"I guess when I'm my dad's age, I'll have black lung, too. Unless something changes, and I don't believe it will."

Shortly after making that statement, Harry Lee quit his job in the mines, but he found that escaping the dust wasn't easy. He could find no other work in the area, and eventually may have to either move north or go back underground.

After four years in the mines, Elmer's nephew, Darrell Mutters, was X-rayed for black lung. Although he had been working primarily on non-production shifts (when less dust is stirred up), the doctor told him the X-ray showed he was fifteen percent disabled. Dust in the mine had to be responsible; he has not smoked a cigarette in his life.

That was five years ago. Today, he's really beginning to feel the impairment of his breath. Unlike Harry Lee Hipshire, Darrell quivers when he talks about it. Grabbing his dinner bucket as he gets ready to leave for midnight shift, he suggests changes he would like to see: new mining technology, dust sampling administered by the miners, fines that would make it too expensive to violate dust standards.

But like Harry Lee, he doesn't expect changes to be made. All he can do, he says through the window of his car as he starts up the engine, is stay on the least dusty shift and take what comes.

His wife, Barbara, stands on the front porch listening, then waves good-night.

"Be careful," she calls after him.

ELMER LOCKHART WITH PAT'S DAUGHTER, CADANCE

God's Country

A rattlesnake buzzes in the distance. Perhaps a coyote has come too close, waking it from a late afternoon nap. Larry Spear reins in his horse to listen. When the rattle stops, there is not a sound on the desert. For dozens of miles in every direction, he sees nothing moving. It is one of the most striking times of day, as the rocks turn golden and pink in the light of the low-lying Arizona sun.

Larry is not sure he is headed in the right direction. He and his wife's father, Boyd Parrish, are supposed to be following opposite sides of an imaginary circle, looking for a wayward cow. But since Larry speaks only English and Boyd only Navajo, Larry was given his instructions primarily with hand signals.

He thinks he is supposed to be staying close to the base of Skeleton Mesa—the long, flat-topped mountain that forms one desert wall—until he comes to the cleared area near some big trees where corn is planted. If he has misunderstood, little harm will be done. His father-in-law will probably find the cow anyway, and the search provides a good excuse for a peaceful ride.

It is the same peace he used to find at this time of day when he worked on merchant marine ships, looking over the rail at the unbroken blue of the Pacific Ocean. As the shadows fall on Black Mesa, the dark plateau across the valley, he leaves behind the pressures of his job at the coal mine on the mesa's surface. He puts out of his mind the breakdown of the dragline earlier in the day that had the whole crew scurrying for tools and parts, knowing that the company was losing five thousand dollars for every hour of downtime.

In this light, he sees Black Mesa for its beauty, as Boyd Parrish sees it. If Boyd had his choice, he would probably have the mine closed down. The strip mining is ruining grazing land, destroying sacred ceremonial areas, and digging up graves. He thinks the white man should not be allowed to treat the land this way.

Larry and his wife, Pearl, understand how he feels. They have recently moved from a trailer in the tiny outpost of Kayenta to an A-frame cabin they built near her father's "hogan," or mud hut. They now live eight miles into the desert on a dirt road, one hundred forty miles from any real towns.

They have no phone, electricity, or running water, but they have other blessings. They are thankful for the land's beauty and quiet, the space for their children to explore, the chance to share her father's knowledge of the desert. And they are thankful, too, that the mining from which they make their living is being done on the mesa across the valley, and not on their land.

NAVAJO MEDICINE MAN M. JOHN

When Larry first came to Navajo country with his new wife in 1970, he brought with him beliefs that did not square well with the traditional customs he found. To him, an accident at the mine or around the house was just that—an accident, an event unrelated to any other. If he got a stomach ache, he saw it strictly as a physical problem, caused perhaps by eating spoiled food or straining a muscle. Nature was just what it appeared to be—rocks, trees, and sand, all there to be used and discarded as necessary by man.

Pearl's father, however, had other ideas. When he looked at the world, he saw Oneness—between body and mind; nature, people and gods; one event and another. His life as a shepherd was a hard one, with little extra margin of comfort. If he suddenly came down with an illness that made it impossible to tend his flock, or if an animal died, he suffered a serious blow, one that could not be shrugged off with, "Oh well, these things happen." *Why* do they happen, he needed to know. Maybe he had violated a taboo, such as talking about certain legends at improper times. If so, if natural disasters were caused by something he had done, then there was hope: perhaps he could prevent future calamities by not making those mistakes again.

He was totally dependent on nature's good will, and in turn, he believed he had to treat his environment with respect if it were to provide for him. So he prayed to nature's spirits, to the sky, the sun and the mesas, asking for rain or a good corn harvest, and he was careful not to waste valuable water and wood. He saw the union of male and female as the source of life: male and female creeks joined to form larger streams, Black Mesa and Skeleton Mesa together formed the desert valley in which he lived. To strip mine tens of thousands of acres of vegetation on Black Mesa was to tear out the mesa spirit's heart, so that it would no longer want or be able to support Navajo life.

Larry didn't mock the Navajo beliefs, but he didn't take them very seriously either, until, in a crisis, he saw them work for his family.

It happened when Pearl was pregnant with their second child. She had been following the Navajo rules for pregnant women, such as not looking at dead animals and not thinking morbid thoughts. But when the baby's due date came, nothing happened. After a week passed, Larry took her to a hospital eighty miles away in the Navajo settlement of Tuba City. The doctor there induced labor, and she went through labor pains for eight hours, but still the baby didn't come, and he sent her home. After two more weeks, she went back to the hospital, where the doctor said she would just have to wait: it was a small baby, and had to grow some more.

Finally, Pearl's sister, Laverne, brought her home from the hospital to see a medicine man. When Pearl told him the baby was overdue, he immediately made the assumption, based on traditional Navajo beliefs, that she or Larry must have tied up an animal for a long period during her pregnancy. Pearl remembered that, in her fifth or sixth month, Larry had left one of the horses tied to a tree for three days.

"That," said the medicine man, "is the problem. We will have to do a ceremony called 'Undo The Rope.'"

He gave her some herbs and pollens he had collected from nearby plants, and began praying and massaging her stomach. While the ceremony was being conducted, her waters broke. The medicine man's wife offered to deliver the baby, but Larry was already outside the hogan with his truck, shouting, "Get in! Get in!" They drove back to Tuba City at ninety miles an hour, and had the baby.

"I'm not going to say what it was that the medicine man did to put her in labor; there could be a lot of explanations for it," Larry says. "But I have to admit that he knew what to do and the doctor didn't.

"I look at my father-in-law, and I realize these people know something about how to live on this land. He's seventy-two years old and I think he could probably outwork any one of us today. He takes his herbs and says his prayers and has the regular ceremonies each year and so forth.

"Most people would say I'm more educated than he is because I've got the book learning over him. But then you take the way the land works, and a way to survive out here, and he's got it over me all the way. Once you know the land and you know the people, you can't laugh it all off anymore."

Pearl Spear's people have lived in northern Arizona and New Mexico for at least a thousand years. For some fifteen generations, they have herded sheep on the desert floor and on the mesas—the plateaus named with the Spanish word for table.

Before the Anglos (white Americans) came, the Navajos had a relatively stable culture. In addition to their animals, they raised corn and other food, and made the baskets and tools they needed. Depending on the time of year, they lived in domed mud hogans or in lean-tos made of branches. Most adults were familiar with herbs that could be used to cure the sick, and complex ceremonies were handed down from one generation of medicine men to the next.

The Navajos were not saints; frequently, and with great skill, they raided the homes of other tribes in the region. Pearl's great-grandfather had seven slaves from the Ute tribe.

"They took women and babies, and horses and cattle to go with them," Pearl says. "The tribes all raided each other. They thought that the more people you had, the larger area you could control."

But then the Navajos met the most powerful raider of all: the U.S. Army. In 1846, the federal government began a campaign to subdue the Navajos which was finally completed in 1864 by Colonel Kit Carson. Carson's men destroyed all crops, livestock and people in their path, and sent messages to those who had not yet surrendered, demanding that they give themselves up at an army post called Fort Defiance.

A few of Pearl's relatives managed to hide on Black Mesa, but most joined the eight thousand Navajos who were taken into custody and then forced to walk three hundred miles to another fort in New Mexico. They were kept in pens, hungry and sick, their spirits broken.

Pearl still speaks of the event. The "Long Walk" is as unforgettable to the Navajos as slavery is to American blacks.

"Our people were smart to go along with it when they saw they were beaten," she says. "Otherwise, we would have been wiped out completely."

When the Navajos were returned to their land, they were under the control of the U.S. government and at the mercy of the Anglo traders. Unfamiliar with arithmetic, they were cheated shamelessly, and were given worthless trinkets for valuable wool, handicrafts, and turquoise. With the Anglos came alcoholism and other new diseases. For decades, the Navajos were encouraged by the government to exploit their main economic resource—sheep—without regard for long-range planning. When overgrazing of sheep began to aggravate erosion of the land and to cut down the sheep's food supply, the Navajos would not reduce their flocks, in part because they saw no way to replace the income they would lose.

With a large family and little food, Pearl's grandparents put her father out on his own when he was nine years old.

"He was weaned like a puppy," Pearl says. "'Here's your sheep,' they told him. 'Take care of them, and you'll have enough food.' The only other thing he had to eat was whatever he could catch, like rabbits."

After Pearl's parents were married, they got a job with a trader, herding sheep on the land the family now occupies. When the government set up a conservation system limiting the number of sheep per acre and restricting use of the land to the Navajos themselves, the Parrishes got the grazing permit for the trader's land.

As a toddler, Pearl heard her people's ancient taboos: don't kill rattlesnakes unless they come near the hogan, don't marry another member of the Red-Streak-In-The-Water clan, don't go out in lightning. She learned about witchcraft—how certain people might try to put a spell on her or her family's animals. To this day, after she combs her hair she gathers up any loose strands and burns them so no one will be able to use them in bewitching her.

Her father also taught her about the white man.

"The white man is always looking for ways to take more from us," he told her. "Ever since the Long Walk he has been telling us what to do."

BOYD PARRISH

When Pearl was six years old, most of her Navajo education ended. She was sent to a series of government boarding schools, where she was taught the white man's ways. Except for vacations, she was to be separated from her family and homeland for the next fifteen years.

The first thing Pearl learned at boarding school was that she was not to speak a word of Navajo. English was taught using books that showed fair-haired Dick and Jane mowing their lawn or going to a baseball game with their white daddy and mommy. The teachers were usually urban blacks, recruited through a variety of government programs. Pearl remembers how confused and embarrassed she felt, standing in front of the class to sing the white man's favorite folk songs, or practicing square dances.

When she reached the third grade, Pearl was sent to another school in Flagstaff, Arizona, where restrictions against speaking Navajo were so tightly enforced, even in the dormitories, that those children who spoke little English rarely talked at all, leading some of the teachers to assume they must not be very bright. The punishment for speaking Navajo, as Pearl learned when someone tattled on her, was to sit alone for two hours, darning other students' socks.

TRAILER CAMP, KAYENTA, ARIZONA

Instead of being taught more about her own religion, Pearl was required to attend church.

''Being out there in the world alone, with no family close you can go to, if you have trouble you're too little to handle it yourself,'' she recalls. ''So after you start going to church, you start believing in the priest, just to have someone.''

For high school, Pearl was moved again, this time to Phoenix, where she was given vocational training to become a dental assistant. That led to a government grant to study dental hygiene at Laney College in Oakland, California. At a friend's house there, she met Larry Spear.

Raised in the state of Washington, Larry is a big man with a strong devotion to work and an equally strong dislike of arbitrary authority. He started sailing on the Columbia River while in high school, and then working for shipping companies transporting supplies for the military during the height of the Vietnam War.

He was later drafted into the Navy, where he had a number of clashes with his bosses, ranging from major issues—for three years, he was unable to get raises in pay and rank to correspond to the work he was doing—to minor harassment. Two weeks before his hitch was up, he began to let his sideburns grow so he would look like a civilian again. The day before his discharge, his commanding officer forced him to shave them off.

''I found out the Navy was like a lot of places,'' he says. ''You had to fight for everything you got, and if they could find a way to get you, they would.''

Several months after Larry and Pearl met, they were married in a San Francisco court. After brief stays in other parts of the country, Pearl persuaded him to try living on the Navajo reservation.

''I always had it in my mind to drag him back here,'' she says. ''As soon as I heard Peabody Coal Company was opening a mine, I figured he could get a job, and I was ready to go. I got some good things from being out in the world, but the Navajo was still the biggest part of me.''

Kayenta was just starting to experience the coal-town boom that is now common in many parts of the West. A Holiday Inn sprang up, then a laundromat, then a bank. For Anglo miners, who are not allowed to occupy Navajo land, and for Navajos who wanted or had to live in town, a trailer park was established, cramming more than a hundred families into an area the size of a Little League baseball field. The Spears, because of Pearl's family ties, were able to move to the country once they built their A-frame.

''I have room out here, I have my freedom,'' Larry says, as he spreads hay for his horses' evening feeding. ''I don't think I could make it if I was stuck in the trailer park, with no recreation except maybe a movie on Thursday nights and a band on Fridays. Sit and drink beer and watch football games on TV, that's about all there is to do down there.''

With no electricity, the Spears have no TV, and their kids don't seem to miss it. Instead, their six-year-old son, Shon, and his cousin, Ervin, help take care of the sheep, horses, chickens, turkeys, and cattle. Three-year-old Rachelle scampers through the brush, copying the way the big kids scoop rainwater from hollowed-out places in the rocks.

For most of the year, Pearl works during the day at the elementary school in Kayenta, where she is a valuable teacher's aide because both Navajo and English are now used in the classroom.

''It's hard for the kids to learn because the languages are so different,'' she says. ''You might not have one Navajo word for every English word; you might have to say something in a completely different way. I got interested in it because I wanted to help out kids in the same situation I was in when I first went to school.''

Pearl works because she enjoys it, not because the Spears need the income. In 1977, by working eighty or more hours per week, Larry made forty-two thousand dollars at the mine. Like other miners, he has found irresistible the overtime rates Peabody pays to avoid having twenty-million-dollar draglines sitting idle. His long hours have won him the respect of Pearl's father, who calls him The Man Who Works All The Time, and the money has financed two new four-wheel-drive pick-ups. But there is little else to spend it on, and now Larry talks about cutting back.

''Now that we're established out here at the cabin, not living in town anymore, I've got my own work I want to do at home,'' he says.

"Why should I work eight hours running a dragline, and then another eight hours running a dozer, when all I'm doing is turning a third of it over to Uncle Sam anyway?"

With the A-frame built, Larry would like to build another house up on the side of the mesa itself, where there is no road, where miners from the trailer park could never bring their dune buggies and jeeps, where nothing spoils the desert's natural state.

"Back up in there," he says, "that's the real God's country. That's where you can *really* get away."

Climbing from the desert floor onto Black Mesa and then approaching the mines eight miles away, the first thing one sees is the dust. The sand on the unmined areas of the mesa is held down by the junipers, greasewoods, pinyon trees, and sagebrush, even in strong summer winds. But those anchors are gone from the thousands of acres which have been mined, resulting in driving dust storms that swirl over the barren land.

Peabody Coal Company is supposed to be reclaiming the mined-out areas, but in the first nine years after the mines opened in 1969, all that grew back were a few scattered, worthless tumbleweeds. Particularly since the average rainfall is less than ten inches per year, real reclamation would cost Peabody money—thousands of dollars per acre, in fact. So the company set up a little charade: it was not really stalling

on reclamation, it was "experimenting." Bulldozers smoothed out the mounds of dirt one way over here, another way over there, adding ruts and ditches in a third place. One type of grass seed was thrown on this patch, another on a second patch, and so on. Meanwhile, the mesa's grazing areas were being turned into an immense, sloped wasteland.

In 1977, Congress passed a new strip mine control law, which set stricter standards for reclamation. Peabody now has a program to set aside the top soil and then put it back in place after the mining is completed. Although the UMWA opposed passage of the law for fear that it would eliminate jobs, predictions by environmentalists that better reclamation would actually create new jobs have come true at Black Mesa. It will not be clear for some time, however, whether even the new reclamation efforts will work in the desert climate.

"If I was Peabody, I wouldn't want to do reclamation either," Larry says. "It's just money out of your pocket that doesn't contribute to your business in any way. The only way you can have reclamation is for somebody to come in and make the company do it."

Approaching the strip pits, one can begin to hear the sounds of mining: blasting, truck engines, and warning sirens and horns. Draglines that stand ten or more stories high scoop out the earth to a depth of about one hundred feet, using a shovel that can hold more than two hundred tons in one bite. The coal that is exposed may be up to eighty feet thick—up to thirty times the thickness of seams in eastern underground mines. Dynamite is used to break up the coal so it can be loaded into gigantic trucks, whose tires alone may be more than ten feet tall.

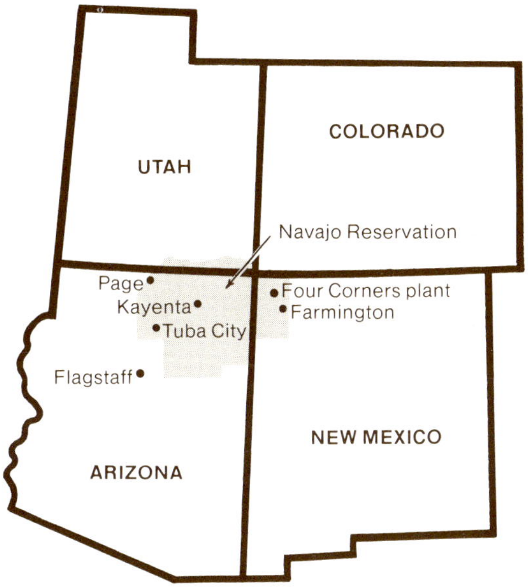

The coal from one of Black Mesa's mines is taken by train seventy miles to a power plant at Page, Arizona, while coal from the other is crushed, mixed with precious water from underground wells, and pumped through a slurry pipeline two hundred forty miles to the Mojave Power Plant in Nevada. The electricity generated at both operations is sent to fuel the resort cities of southern Arizona, and to bring heat and light to the factories, homes and swimming pools of Los Angeles.

When the wind blows in one direction, sulphur-filled smoke from coal burned at the Page plant carries all the way to Kayenta, so thick that the setting sun cannot be seen behind it; on other days, smoke comes from the Four Corners power plant near Farmington, New Mexico, which dumps over one hundred million pounds of sulphur into the air each year. Concerned about the pollution as a health hazard, the Navajo tribe is trying to impose stiff fines to force the utilities to install sulphur-control devices.

But the most immediate danger to the Black Mesa miners is right on the job. Already, a number of Navajos have been killed in accidents at the mines, and bulletin boards in the Kayenta trading posts, which once mainly carried notices to tourists, now periodically announce the one-day memorial periods which are traditional in the coal industry after a miner's death.

The mine is a jungle of the fast-moving steel cables, gears and belts, high-voltage wires, and massive machinery that make some strip mines even more dangerous than underground operations. On the mine's long, winding access road, low visibility caused by dust is a special hazard. In June, 1977, a strike was called to demand that the road be dampened with oily soap instead of pure water that evaporates within minutes. Peabody threatened the miners with a lawsuit, and the next day they were back on the job without any improvements. Exactly one week

later, a miner who apparently thought he had a clear way ahead of his car took his eyes off the road and collided at sixty miles an hour with the back of a truck.

When another worker was crushed under a conveyor belt several years earlier, federal inspectors found three major violations of the law: failure to properly train the miner; allowing the miner, who had only two months of mining experience, to work alone; and failure to keep the steel guard over the belt permanently bolted down. The young Navajo had been assigned to clean up coal spillage around the belt, and apparently thought he was supposed to clean under it as well. The mine superintendent at the time, a boss imported by Peabody from Kansas, told two UMWA representatives that lack of proper training and supervision had not been the problem at all. Instead, the accident was just more evidence that Navajos were too dumb to be coal miners.

"Most of them are pretty dense," he complained openly. "They can't speak, they can't read or write. If they would just adopt more of the white ways it would be so much easier. They won't because they don't trust the white man, but I don't see why not."

After the superintendent's remarks were reported in the UMW Journal, Navajos in the local union successfully demanded that he be transferred back to the Midwest. They were supported by the Anglo miners, who make up about twenty percent of the Black Mesa work force of seven hundred. While a few of the Anglos, like Larry, are married to Navajos, most of them were brought in by Peabody to help start the mine and train Navajo miners. Because they have past experience in dealing with the UMWA and with coal company managers, Anglos have been elected by the miners to most of the key local union offices. But they have opposed discrimination by the company, and are training Navajos to play a larger role in the union.

"We know that without the UMWA, Peabody would have paid us low wages like people have always been paid on the reservation," says Roy Tutt, one of an increasing number of Navajo local union officers.

"At first, the company would treat the Anglos different, like putting them in charge of a crew even when there was a more experienced Navajo there, but we brought it up through the union and it doesn't happen too much anymore."

As the Black Mesa miners became familiar with their union, they began to wonder why visits to medicine men weren't covered by the UMWA Health and Retirement Funds, the way doctor's visits are. If the Health Funds would pay even part of the cost, perhaps more miners would go to medicine men, and an important part of Navajo culture would have a better chance to survive. At the UMWA international convention in 1976, the Black Mesa delegates, including Larry Spear, Roy Tutt, and a Navajo named Charles Harris, decided to raise the issue.

They were uncertain how their resolution would be received. The level of tolerance on the convention floor had fluctuated wildly; at one point, the miners gave a spontaneous round of applause for the first woman miner ever to speak as a UMWA convention delegate, but at another point they interrupted debate on mine safety resolutions for an hour to consider impassioned pleas for a search of the elected delegates to root out "communists."

(FOREGROUND) ROY TUTT AND LARRY SPEAR AT UMWA CONVENTION

"I thought we might get laughed off the floor, talking about something like medicine men," Larry says, "and there was one or two that made comments. But most of the guys showed a lot of respect for our problems."

The convention voted to endorse Health Funds coverage for medicine men, and in local contract negotiations with Peabody in late 1977, the company agreed to the proposal.

But despite that success, the UMWA is on shaky footing at Black Mesa. Before the 1977 contract negotiations, a smaller, more company-oriented union, the Progressive Mine Workers, forced an election to determine whether the Black Mesa miners wanted to switch unions. The UMWA won the vote, but it is likely that the Progressives or some other union will try again before the next round of contract bargaining.

Like other western miners, the Black Mesa workers have little sentimental attachment to any particular union; they do not have the memories an eastern miner might have of getting a new pair of shoes after the UMWA won a raise for his father or grandfather. They are in the coal industry for the money, and another union which can offer more could easily win their loyalties.

"More dollars on the day, that's the bottom line," says Larry Spear. "If somebody comes in with an offer, I'm afraid a lot of people here will listen."

It could be to Peabody's advantage to help some other union, even by paying higher wages or bonuses based on productivity from the thick seams, if it could negotiate smaller medical and pension payments, fewer job health and safety protections, and a ban on strikes. A similar strategy has already been followed at most of the mines which have opened in the West in recent years. Western production jumped five hundred percent from 1970 to 1976, with virtually all of the new operations represented by construction unions, company unions, or no union at all. The largest mine on Navajo land, run by Utah International to supply the Four Corners plant, is under contract with the Operating Engineers, AFL-CIO.

Because of the new non-UMWA mines in the West, coal produced by UMWA members was only about fifty-five percent of the national total in 1977, compared to about seventy percent seven years earlier. Eventually, the industry may be in a position to refuse to renew or improve contracts at UMWA mines, and then to use non-UMWA western coal to fill customers' orders and maintain profits during a national UMWA contract strike.

Peabody, which was purchased in 1977 for more than one billion dollars by a group of giant companies, including the Boeing Corporation and Equitable Life Assurance, plans several more huge mines in the West. Continental Oil, Occidental Petroleum. Standard Oil of California, Exxon, Gulf Oil, and Shell all have purchased millions of tons of western reserves. Corporations such as these plan another tripling of western coal production by 1985. By that year, more coal may be mined in Wyoming than in West Virginia.

''What happens at a mine like ours is going to determine whether an eastern coal miner even has a union five or ten years from now, maybe not that long,'' Larry says. ''It's kind of strange when you think about it, but this is where the future of the East is being decided, right out here.''

As coal development has expanded on Navajo land, opposition to it on environmental or religious grounds has received national publicity. The issue has always been described as ''jobs versus environment,'' or ''progress versus tradition.'' But another kind of criticism has been ignored: some Navajos who do not work at the mines believe that the development will not even provide long-term progress for the majority of Navajos.

The tribe might have been able to operate the mines on its own by hiring experienced managers and technical advisors. Then it would have kept most of the profits, and could have carried out proper reclamation. Instead, the tribe signed a lease with Peabody that provided only twenty-five cents per ton for coal that sells for more than six dollars per ton. Little of the money the tribe has made on coal leases during the past fifteen years has been invested in other types of businesses to create more jobs. As a result, the unemployment rate on the reservation remains near fifty percent. Proposed new mines and coal gasification plants will cut that rate by only about ten percent.

The reservation, like other western communities, could face a fate similar to Appalachia's. Its coal riches will be extracted by outside interests, leaving behind little other industry, inadequate community services, and land that is no longer useful for any other purpose. While a few thousand Navajos will prosper temporarily, tens of thousands will still be without jobs and will lose grazing lands for their sheep. In thirty-five to forty years, when the coal will be mined out, Kayenta could become a ghost town of abandoned trailers.

Despite the UMWA's long experience with this cycle in the eastern coalfields, neither the international nor local unions have supported groups raising these questions. In fact, the Black Mesa miners are uncomfortable when they hear any criticism of the current mining operations.

''If my father-in-law's land was being torn up where he herds his sheep and has certain sacred areas, he wouldn't like it, and I don't guess I would like it either,'' Larry says. ''I know if I put my loved one into the ground I would hate to see some dragline come along and dig up that

CHARLES HARRIS

grave. But that's just the way the world goes—somebody loses out. If you start talking about preserving the traditions, when you start pointing out how they're not doing their reclamation, then pretty soon you'll have those radicals in here wanting to shut us down all the way."

Asked about the success of reclamation at the mines, Navajo miner Roy Tutt squirms in his chair.

"I think the land could be put back the way it was, so it could still be used for grazing or something else," he says.

Is it being put back?

"I think it could be, if the reclamation is done right, yes."

But *is* the reclamation being done right?

"If they plant what they're supposed to and grade it and so on, I think it could be done.

"The important thing," Tutt says, "is that we've got jobs. Before, there weren't any jobs for our people at all. We're making good money. This is good for us. So I'm in favor of what they're doing."

Charles Harris says he would like to see better reclamation, but "you've got to have progress."

"We can't be left behind," he says. "We've got to stay up there and hang in the race."

It is obvious, however, that many of the miners still feel ambivalent about the stripping. According to Pearl Spear, some of them go to medicine men who believe that illnesses and accidents they suffer are caused by the gods in retaliation for the destruction of Black Mesa.

"Forgive this man," the medicine man will pray as he conducts a ceremony over the sick or injured miner.

"Make him well again. He knows it is wrong to destroy Black Mesa. But forgive him. He must eat, and this is the only way he has to make his living."

Matt Witt (left) and Earl Dotter (right) joined the staff of the United Mine Workers Journal in 1973 after helping rank-and-file miners oust former UMW President Tony Boyle. Witt served as the Journal's managing editor and then editor, and Dotter was its photographer, until both resigned in November, 1977. In 1976, the Journal became the first labor union publication to win the National Magazine Award sponsored by the American Society of Magazine Editors and Columbia School of Journalism. The Journal was named the best American magazine in the field of specialized journalism. In 1977, Witt and Dotter were among the five finalists chosen by the award's sponsors in the category of reporting excellence. Their work has appeared in a number of other magazines, newspapers, and labor union publications.

Photographs for this book were provided by Walter Rosenblum (page 6), Eastman House (page 6), and Jay Chatman (page 23 and 26).

Design by Charles Myers
Production by Karen Ohmans